SpringerBriefs in Applied Sciences and Technology

SpringerBriefs present concise summaries of cutting-edge research and practical applications across a wide spectrum of fields. Featuring compact volumes of 50 to 125 pages, the series covers a range of content from professional to academic.

Typical publications can be:

- A timely report of state-of-the art methods
- An introduction to or a manual for the application of mathematical or computer techniques
- A bridge between new research results, as published in journal articles
- A snapshot of a hot or emerging topic
- An in-depth case study
- A presentation of core concepts that students must understand in order to make independent contributions

SpringerBriefs are characterized by fast, global electronic dissemination, standard publishing contracts, standardized manuscript preparation and formatting guidelines, and expedited production schedules.

On the one hand, **SpringerBriefs in Applied Sciences and Technology** are devoted to the publication of fundamentals and applications within the different classical engineering disciplines as well as in interdisciplinary fields that recently emerged between these areas. On the other hand, as the boundary separating fundamental research and applied technology is more and more dissolving, this series is particularly open to trans-disciplinary topics between fundamental science and engineering.

Indexed by EI-Compendex, SCOPUS and Springerlink.

Ahmad Amiri · Hossein Shahali ·
Andreas A. Polycarpou

The Zinc–Sulfur Battery

The Next Frontier in Energy Storage
Technology

 Springer

Ahmad Amiri
Russell School of Chemical Engineering
The University of Tulsa
Tulsa, OK, USA

Department of Mechanical Engineering
The University of Tulsa
Tulsa, OK, USA

Andreas A. Polycarpou
Department of Mechanical Engineering
The University of Tulsa
Tulsa, OK, USA

Hossein Shahali
Russell School of Chemical Engineering
The University of Tulsa
Tulsa, OK, USA

ISSN 2191-530X   ISSN 2191-5318   (electronic)
SpringerBriefs in Applied Sciences and Technology
ISBN 978-3-031-71490-0   ISBN 978-3-031-71491-7   (eBook)
https://doi.org/10.1007/978-3-031-71491-7

This Springer imprint is published by the registered company Springer Nature Switzerland AG
The registered company address is: Gewerbestrasse 11, 6330 Cham, Switzerland

If disposing of this product, please recycle the paper.

Contents

Abbreviations

1D CNFs	One-dimensional carbon nanofibers
2DZS	Two-dimensional zinc/sulfur
AC-S	Sulfur-added activated carbon
ACSA	Acid-assisted confined self-assembly
AE	De-ionized water
AEAs	All-electric airplanes
Al-S	Aluminum-sulfur
AN	Acetonitrile
AZIBs	Aqueous zinc-ion batteries
BES	Battery energy storage
CA	Chronoamperometry technique
CE	Coulombic efficiency
ChCl	Choline chloride
CMK-3	Mesoporous carbon material
CNF-S	Carbon nanofibers enriched with sulfur
CO_2	Carbon dioxide
CoS	Cobalt sulfide
CV	Cyclic voltammetry
DES	Deep eutectic solvents
DFT	Density functional theory
DMF	Dimethylformamide
E-cycle	Electrochemical cycles
EDS	Energy-Dispersive X-ray Spectroscopy
EG	Ethylene glycol
EIS	Electrochemical impedance spectroscopy
EPS	Electrochemical power sources
ESPW	Electrochemically stable potential window
ESSs	Energy storage systems
EVs	Electrical vehicles
FTIR	Fourier Transform Infrared
G_4	Tetraglyme

GCD	Galvanostatic charge/discharge
GITT	Galvanostatic Intermittent Titration Techniques
H_2S	Hydrogen sulfide
HAADF-STEM	High-Angle Annular Dark Field-Scanning Transmission Electron Microscopy
HCS	Hollow carbon spheres
HER	Hydrogen evolution reaction
I	Iodine
I_2	Di-iodine
ILs	Ionic liquids
In	Indium
KJS	Kruk-Jaroniec-Sayari
LF	Liquid film
LIBs	Lithium-ion batteries
$LiPF_6$	Lithium hexafluorophosphate
Li-S	Lithium-sulfur
LSV	Linear sweep voltammetry
LUMO	Lowest unoccupied molecular orbitals
$Me_3PhN^+I^-$	Trimethylphenylammonium iodide
N/P	Negative-to-positive
Na-S	Sodium-sulfur
Ni-MH	Nickel-metal hydride
NMP	Methyl-2-pyrrolidinone
NPC	Nonporous carbon
OER	Oxygen evolution reaction
OTF^-	Trifluoromethane sulfonate anions
PAN	Polyacrylonitrile
PLSD	Poly $Li_2S_6 - r - DIB$
PMMA	Polymethylmethacrylate
PPE	Personal protective equipment
PSD	Pore size distribution
PTFE	Polytetrafluoroethylene
PVDF	Polyvinylidene fluoride
PVP	Poly (N-vinyl-2-pyrrolidone)
pZn/In	Powder zinc/indium
S@C	Sulfur embedded in carbon
S@CNF	Sulfur embedded in carbon nanofiber
S@CNTs	Sulfur embedded in carbon nanotubes
S@Fe-PANI	The combination of sulfur with iron cathode
S@KB	Sulfur embedded in Ketjen Black
S@NPC	Carbon-supported sulfur cathode
SABs	Sulfur-aqueous-based batteries
SEAD	Selected area electron diffraction
SEI	Solid electrolyte interphase
SEM	Scanning Electron Microscopy

SHE	Standard hydrogen electrode
SO_2	Sulfur dioxide
SSE	Solid-state electrolyte
TEM	Transmission Electron Microscopy
TEOS	Tetraethyl orthosilicate
TGA	Thermal gravimetric analysis
TU	Thiourea
TU^-	Thiourea ion
TUI	Iodinated thiourea
UAVs	Unmanned aerial Vehicles
W-Zn	1 M $ZnSO_4$ in water
XPS	X-ray Photoelectron Spectroscopy
XRD	X-ray Diffraction
ZAPS	Zinc-aqueous polysulfide
ZIBs	Zinc-ion batteries
ZnI_2	Zinc iodide
Zn–NHPC	Zn–N_4 on nitrogen-doped hollow porous carbon
ZnS	Zinc sulfide
Zn-S	Zinc-sulfur batteries
ZnS@CF	Carbon fiber-reinforced ZnS
ZPBs	Zinc-polysulfide batteries

Chapter 1
Introduction to Zinc–Sulfur Batteries

1.1 Overview of Battery Technology

In the wake of rapid global industrialization, the imperative for electrical power has reached a critical juncture, propelled by the concurrent challenges of energy transformation and deficiency in reliable supply. This urgency stems from the finite reserves of traditional energy sources, which not only contribute significantly to exacerbating global climate change but also play a pivotal role in heightening challenges related to water availability. The release of detrimental chemical pollutants, including sulfur-based and nitrogen-based compounds, further amplifies these threats to both the environment and human well-being [1]. As nations grapple with the consequences of finite fossil resources and their environmental toll, there is an urgent need to explore and adopt cleaner and renewable energy sources. The harnessing of solar, wind, hydropower, biomass, and geothermal energy emerges as a promising avenue for meeting the escalating energy needs while simultaneously mitigating the adverse impacts on the planet [2, 3].

While renewable energy sources hold tremendous promise for addressing environmental concerns and fostering sustainability, they are confronted with notable challenges that warrant consideration. One primary hurdle lies in their intermittent availability, as sources like solar and wind power are subject to variations in weather conditions. This intermittency poses challenges to maintaining a consistent and reliable energy supply, requiring innovative solutions such as energy storage technologies. Instability is another key concern associated with renewable energy. The inherent unpredictability of natural elements, such as fluctuations in wind speed or sunlight intensity, can result in inconsistent power generation. This variability introduces complexities in managing and balancing the electrical grid, necessitating sophisticated grid management systems and resilient infrastructure to accommodate the dynamic nature of renewable energy sources. Furthermore, natural resource limitations present a challenge to the widespread adoption of renewable energy. For instance, the availability of suitable locations for large-scale hydropower or

© The Author(s), under exclusive license to Springer Nature Switzerland AG 2024
A. Amiri et al., *The Zinc–Sulfur Battery*,
SpringerBriefs in Applied Sciences and Technology,
https://doi.org/10.1007/978-3-031-71491-7_1

geothermal projects may be geographically limited. These limitations could result in potential discrepancies between the supply and demand for renewable energy technologies, hindering their seamless integration into existing energy infrastructures. Responding to these hurdles, the promotion of groundbreaking and eco-friendly renewable energy sources has become a pivotal tactic in striving for global progress toward sustainable and low-carbon energy development and provision.

The imperative to address the hurdles associated with renewable energy accentuated the need for advanced energy storage technologies, encompassing both batteries and capacitors. The historical context of over a century of development in electrochemical energy storage attests to the enduring commitment to finding innovative solutions.

Out of all energy storage systems (ESSs), batteries distinguish themselves with their efficiency, convenience, reliability, user-friendly operation, and minimal maintenance requirements [4]. To store electrical energy, it must undergo conversion into another form, and batteries emerge as a prominent and logical choice for such a purpose [5]. Additionally, batteries are poised to play a pivotal role in incorporating intermittent renewable energy sources such as wind and solar power. Their ability to smooth out energy output and enhance the versatility of renewable energy in micro-generation systems enables a consistent supply and distribution of electrical power [6–8]. Setting aside considerations of cost and environmental impact, battery energy storage (BES) stands out as one of the most efficient methods for stabilizing power grids, particularly those heavily reliant on substantial quantities of renewable energy, surpassing the 10% threshold [9]. In the diverse array of BES types, lithium-ion batteries (LIBs) take the lead, commanding an impressive 55% market share. This dominance underscores their widespread adoption and effectiveness in addressing the challenges associated with stabilizing power grids and optimizing the utilization of renewable energy sources [8].

LIBs employing intercalation chemistries are acknowledged as the foremost battery technology, given their superior qualities in energy density, efficiency, and longevity over alternative types of batteries. LIBs excel in powering small-scale electronics and find extensive applications in renewable energy and micro-grid systems [8]. Their appeal lies in a range of advantages, encompassing sealed cells requiring minimal maintenance, a prolonged cycle life, operational viability across a wide temperature spectrum, swift charging capabilities, high efficiency in charge and discharge cycles, impressive energy density, and substantial design adaptability [10]. The design flexibility extends to the choice of salts for the electrolyte, conventionally featuring lithium hexafluorophosphate ($LiPF_6$) [11]. However, LIBs employing $LiPF_6$ face challenges related to thermal stability, moisture sensitivity, and the release of toxic by-products [12]. Another avenue for enhancing LIB efficiency involves incorporating solid-state electrolytes due to their superior thermal and chemical stability [13]. Nevertheless, a rising apprehension concerning their expense, safety issues, the restricted accessibility of lithium and other raw materials, and the negative environmental consequences resulting from heightened usage have instigated a quest for alternative battery technologies [14–16]. These technologies involve utilizing plentiful metals as the primary materials and employing eco-friendly and non-combustible electrolytes, all while showcasing robust electrochemical performance [3, 17, 18]. Over the past few years, there has been a pronounced emphasis on investigating zinc-ion batteries (ZIBs) as a viable alternative, attracting considerable interest among researchers in the battery technology domain [19–21]. The attractiveness of zinc to these researchers stems from its commendable traits, encompassing stability, safety, volumetric capacity, reversibility in water-based settings, and economical nature. The favorable qualities of zinc batteries comprise:

i. Incorporation of plentiful elements: In the design of ZIBs, easily accessible components are utilized, encompassing zinc metal for the anode, aqueous zinc salt solutions for the electrolyte, and metal oxides for the cathodes. Zinc, standing as the 24th most prevalent element in the Earth's crust, is not only economically feasible (priced at around $2 per kg) but also abundant (reaching approximately 75 ppm) [22–26].
ii. Safety and environmental suitability: In contrast to other metals employed in batteries such as lead, lithium, magnesium, and cadmium, zinc stands out for its non-toxic properties and poses no safety hazards. Furthermore, the ability to employ aqueous electrolytes is facilitated by the redox potential of zinc, which is measured at $- 0.763$ V compared to the standard hydrogen electrode [3].
iii. Environmentally friendly non-combustible electrolytes: The intrinsic characteristics of zinc enable its compatibility with a diverse array of solid, gel, and liquid electrolytes. This aspect has spurred substantial research endeavors focused on ZIBs.

Enhancing the electrochemical performance, a pivotal benchmark in the energy storage sector entails the identification of multiple suitable cathode materials. Examples of these materials encompass MnO_2, phosphate, vanadates, Prussian blue, and organic compounds [27–31]. Nevertheless, challenges arise for these cathodes, presenting issues like material dissolution and detrimental interactions between Zn^{2+} ions and structures of the host materials. Despite the considerable theoretical capacity of the zinc anode (820 mAh/g), the capacity of intercalation-based cathodes usually falls within the range of 50–300 mAh/g, thereby restricting the overall energy density of ZIBs. However, the utilization of conversion mechanisms in cathodes holds the promise of improving energy density, potentially addressing this challenge [32, 33]. This idea has spurred widespread research into metal-sulfur batteries utilizing organic electrolytes, such as lithium–sulfur (Li–S) and sodium-sulfur (Na–S) batteries [32]. Although sulfur has commonalities with zinc concerning its abundance, affordability, eco-friendliness, and considerable theoretical capacity of 1675 mAh/g, metal-sulfur batteries comprised of organic electrolytes face challenges akin to intercalation-based LIBs, particularly concerning safety and sustainability issues. Furthermore, issues associated with the migration of polysulfide compounds impede battery durability and coulombic efficiency (CE). The amalgamation of zinc and sulfur gives rise to Zinc–Sulfur (Zn–S) batteries, representing an eco-friendly and economical energy storage technology with a significant energy density surpassing 500 Wh/kg when compared to current alternatives (Fig. 1.1). Figure 1.1 provides a comprehensive graphical comparison of the specific power and energy density characteristics of

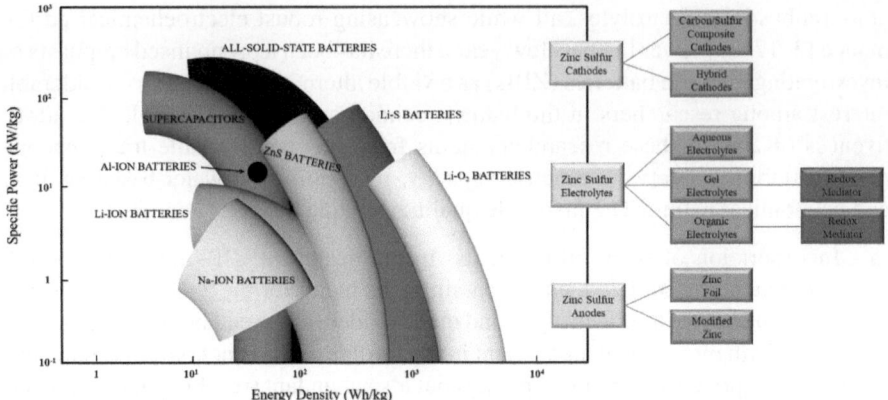

Fig. 1.1   Ragone plot illustrating the exceptional electrochemical performance of Zn–S batteries, solidifying their position as formidable rivals to LIBs. Additionally, the improvement of various electrolytes, as well as cathode and anode materials, was delved into, and they were systematically organized into categories. Reproduced with permission from [34]

various prominent and widely used ESSs, including supercapacitors and batteries. By illustrating the Ragone plot of these ESSs, it elucidates their status as a function of power density versus energy density.

1.2   Historical Background

The schematic representation in Fig. 1.2 not only provides a visual insight but serves as a portal into the captivating historical journey of Zn–S batteries. Dating back to 1836, the inception of Zn–S batteries is a testament to the visionary efforts of pioneers who embarked on the exploration of ZIB development, laying the groundwork for a trajectory that would unfold over two centuries.

Fast forward to 2002, a watershed moment unfolded as scientists delved into the intricate thermodynamics of Zn–S cells, unraveling complexities that paved the way for a theoretical framework governing the dynamics of Zn–S systems. This breakthrough set the stage for subsequent advancements, propelling the technology forward [35].

The transformative years of 2018 and 2020 marked a turning point with the advent of a reversible zinc-aqueous polysulfide battery [36] and the introduction of Zn–S batteries [37]. These milestones not only bolstered the energy performance of Zn–S batteries but also ushered in opportunities for sustainable and eco-friendly energy storage alternatives, steering the technology toward a more environmentally conscious future.

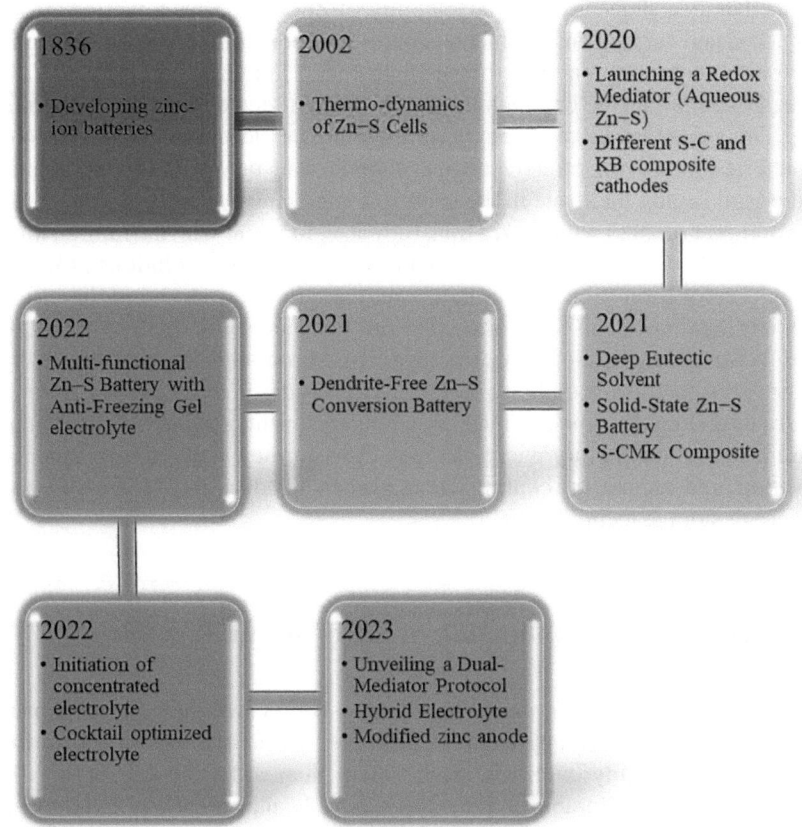

Fig. 1.2 Chronology of Zn–S battery evolution. Reproduced with permission from [34]

During that particular timeframe, the incorporation of a redox mediator into aqueous Zn–S settings emerged as a pivotal development, enhancing stability and efficiency [38]. This innovation shattered preconceived limits, expanding the horizons of what was deemed achievable in Zn–S battery technology.

The groundbreaking leap in 2021, with the introduction of solid-state Zn–S batteries, elevated safety and reliability standards [39]. This marked a significant stride in realizing efficient and scalable Zn–S battery solutions, pushing the boundaries of technological capabilities.

Maintaining the momentum, 2021 witnessed the development of Zn–S batteries resistant to dendrite formation, addressing a critical concern in commercial applications [40]. Surmounting the obstacles linked to the formation of dendrite structures, this advancement bolstered the durability and steadfastness of Zn–S batteries, reinforcing their viability in practical scenarios.

In 2022, a significant breakthrough occurred with the unveiling of a versatile Zn–S battery design [41]. This groundbreaking technology not only enhanced overall performance but also showcased the adaptability of Zn–S batteries to various technological needs and varying environmental conditions, positioning them as dynamic solutions for contemporary energy storage challenges. Building upon prior accomplishments, 2022 marked the commencement of concentrated electrolyte configurations [42]. This inventive strategy sought to fine-tune the electrochemical reactions in Zn–S batteries, improving effectiveness, and broadening the spectrum of potential uses, further solidifying their role in the energy storage landscape.

The pivotal year of 2023 witnessed a transformative moment with the introduction of an innovative dual-mediator approach and hybrid electrolytic solutions [43, 44]. This strategic innovation promised to tackle intricate hurdles, elevating the effectiveness and reliability of Zn–S batteries, and firmly establishing them as frontrunners in the ever-evolving field of energy storage. This chronological narrative vividly portrays the enduring transformation and noteworthy strides in the evolution of Zn–S batteries, transcending their modest inception to their current prominence as cutting-edge solutions in contemporary energy storage. The journey, marked by resilience and innovation, underscores the continual commitment to advancing sustainable and efficient energy storage technologies.

1.3   Importance and Application of Zinc–Sulfur Batteries

Zn–S batteries are rapidly emerging as highly promising candidates for an expansive range of energy storage applications, attributed to their integration of cost-effectiveness, compatibility with air, high safety standards, significant anodic capacity, and prolonged durability [45, 46]. The advantages of Zn–S batteries encompass:

i. In contrast to the prevalent metal-sulfur batteries such as lithium-sulfur, sodium-sulfur, aluminum-sulfur, magnesium-sulfur, and calcium-sulfur, which often grapple with susceptibility to air and are deemed unsuitable for charging and discharging in ambient conditions, Zn–S batteries exhibit a notable departure. One of the distinguishing features of Zn–S batteries is their remarkable compatibility with the surrounding air. This characteristic not only sets them apart from their counterparts but also enhances their versatility and practicality in commercial applications.

ii. The achievement of heightened safety in Zn–S batteries is a pivotal breakthrough facilitated by the strategic incorporation of aqueous electrolytes. This deliberate choice not only enhances the overall performance of the batteries, reaching their theoretical capacity of 820 mAh/g but also serves as a robust safety measure, significantly minimizing the risk of potential issues such as leakage and combustion.

iii. Water's adaptability shines through when crafting specialized electrolytes. By combining water with other solvents or strategically amalgamating it with diverse additives, we can improve the performance of these electrolytes, ushering in a new era of customization and efficiency in energy storage technology.

iv. Environmental harmony is a distinctive feature of Zn–S batteries, attributed to the abundant and eco-friendly nature of their constituent materials, zinc, and sulfur. These elements not only offer plentiful resources but also boast non-toxic properties, resulting in negligible environmental repercussions throughout the entire lifecycle of Zn–S batteries-form production to recycling. This inherent eco-friendliness positions Zn–S batteries as a sustainable and responsible choice in the landscape of energy storage technologies in sharp contrast to some other batteries, such as those utilizing lithium metal, where stringent prerequisites are enforced due to concerns related to toxicity and negative environmental impact.

v. The economic viability of Zn–S batteries is underpinned by the inherent affordability of their primary components: zinc and sulfur. The ample supply and accessibility of these elements ensure a stable and cost-effective foundation for large-scale battery production. This cost-effectiveness positions Zn–S batteries as a financially viable and attractive option for extensive energy storage applications in the foreseeable future.

vi. Durability stands as a defining characteristic of Zn–S batteries, as they showcase an exceptional operational lifespan that allows for multiple cycles of charge and discharge. This robust durability not only ensures sustained performance over time but also underscores the reliability of Zn–S batteries in meeting the demands of diverse energy storage applications.

In a Zn–S battery, akin to ZIBs, the process of charge retention entails the migration of zinc ions across an electrolytic medium. At the sulfur electrode, conversion reactions transpire, exchanging two electrons between the electrodes and yielding a theoretical voltage of 1.15 V [35]. Nevertheless, even with the advantages provided by aqueous electrolytes and organic substitutes, Zn–S batteries encounter considerable challenges [22, 30, 47–51], which include:

1. A significant impediment in the advancement of Zn–S batteries revolves around the restricted utilization of sulfur, stemming from its inherent insulating attributes and sluggish kinetic reactions. Overcoming this challenge necessitates the construction of cathodes with enhanced conductivity and active sites. Attaining this objective requires the inclusion of additives with superior conductivity or the utilization of sophisticated materials such as carbon composites or modified electrolytes. These modifications enhance both electron transfer during the charge and discharge processes and the kinetics of the sulfur reaction [40, 43, 52, 53].

2. The expansion of the cathode during the zincation process represents a critical challenge in the operational life of Zn–S batteries. This expansion induces mechanical impairment to the electrodes, ultimately leading to a significant reduction in its cycling capability. The consequences of this mechanical stress are multifaceted and pose hurdles to the long-term performance and efficiency of the battery system.

3. The diminished redox potential of Zn/Zn^{2+}, measured at -0.762 V versus standard hydrogen electrode (SHE), coupled with the favorable kinetics of the hydrogen evolution reaction (HER) compared to the oxygen evolution reaction (OER), significantly influences the behavior of the zinc anode in Zn–S batteries. This interplay elevates the probability HER occurring on the surface of the zinc anode throughout the cycling of the battery. The persistent occurrence of HER on the surface of the zinc anode results in continual water consumption and irreversible processes of zinc plating and stripping.
4. The considerable diminish in voltage efficiency in Zn–S batteries is attributed to specific operational characteristics, notably a discharge voltage below 0.5 V and significant polarization exceeding 0.8 V. These factors contribute to challenges in maintaining a high voltage efficiency, impacting the overall performance and effectiveness of the battery system.
5. The proneness of zinc metal to corrosion in aqueous electrolytes results in uneven formation of by-products on the anode surface, intensifying dendrite growth and contributing to the deterioration of the electrolyte. Additionally, this corrosive process destabilizes the surface of the anode, jeopardizing the stability of electrode reactions and the overall battery performance.

To sum up, although the Zn–S battery positions itself as a promising substitute for ZIBs in diverse applications, showcasing advantageous attributes such as cost-effectiveness, environmental sustainability, non-flammability, substantial theoretical capacity, enhanced performance, and lightweight design, it is crucial to acknowledge and consider its inherent constraints. Conducting comprehensive exploration and in-depth investigation is essential to grasp the physicochemical traits within diverse electrolyte families. This includes the identification of electrode and electrolyte materials characterized by robust electrochemical attributes and the clarification of interactions and influential mechanisms existing between electrolytes and electrode materials. The overarching aim of this pursuit is to optimize battery performance. As of now, the progressions of Zn–S batteries remain in their preliminary phase. This book provides a comprehensive overview of the research advancements in Zn–S batteries. To accomplish this objective, we have structured recent progress into three distinct classifications: materials for anode, materials for cathode, and the evolution of electrolytes, as depicted in Fig. 1.1. This book places a strong emphasis on scrutinizing the inherent reaction mechanisms occurring during the charge and discharge processes of Zn–S batteries. The investigation extends to the improvement of various electrolytes, covering aqueous and organic formulations, along with gel-based electrolytes. Additionally, this book explores the influence of redox mediators and alternations on cathode materials, examining how they affect the processes of storing charge and kinetic characteristics of Zn–S batteries. Based on this exploration, this book methodically assesses the data representing electrochemical behaviors of improved materials, succinctly outlining both their advantages and drawbacks. Furthermore, this book proposes potential avenues for future investigation and prospects, offering insights to navigate the continued progress of this technology.

References

1. R. Demir-Cakan et al., J. Mater. Chem. A **7**(36), 20519 (2019)
2. B. Li et al., Nano-Micro Lett. **14**(1), 6 (2021)
3. H. Kim et al., Chem. Rev. **114**(23), 11788 (2014)
4. T.B. Reddy, *Linden's Handbook of Batteries* (McGraw-Hill Education, 2011)
5. H.A. Kiehne, *Battery Technology Handbook* (CRC Press, 2003)
6. M.C. McManus, Appl. Energy **93**, 288 (2012)
7. M.A. Rahman et al., J. Electrochem. Soc. **160**(10), A1759 (2013)
8. C. Zhang et al., Renew. Sustain. Energy Rev. **82**, 3091 (2018)
9. G.J. May et al., J. Energy Stor. **15**, 145 (2018)
10. T.B. Reddy (2002)
11. J.B. Goodenough, Y. Kim, Chem. Mater. **22**(3), 587 (2010)
12. A. Mauger et al., Mater. Sci. Eng. R. Rep. **134**, 1 (2018)
13. B. Commarieu et al., Curr. Opin. Electrochem. **9**, 56 (2018)
14. B. Li et al., Nano-Micro Lett. **14**, 1 (2022)
15. T. Jin et al., Chem. Soc. Rev. **49**(8), 2342 (2020)
16. L. Zhang et al., J. Mater. Chem. A **11**(6), 3105 (2023)
17. T. Ould Ely et al., Front. Energy Res. **7**, 71 (2019)
18. J. Liu et al., Green Energy Environ. **3**(1), 20 (2018)
19. J. Yue, L. Suo, Energy Fuels **35**(11), 9228 (2021)
20. S. Ponnada et al., Energy Fuels **36**(12), 6013 (2022)
21. J. Wang et al., Energy Fuels **35**(8), 6483 (2021)
22. C. Xu et al., Angew. Chem. **124**(4), 957 (2012)
23. W. Manalastas Jr. et al., Chemsuschem **12**(2), 379 (2019)
24. C. Xie et al., Energy Environ. Sci. **12**(6), 1834 (2019)
25. J. Zhang et al., Chem. Sci. **10**(39), 8924 (2019)
26. Y. Li, J. Lu, ACS Energy Lett. **2**(6), 1370 (2017)
27. W. Shi et al., Chemsuschem **14**(7), 1634 (2021)
28. L. Ou et al., Chemsuschem **15**(19), e202201184 (2022)
29. X. Chen et al., Energy Storage Mater. **50**, 21 (2022)
30. Y. Li et al., Mater. Today Energy, 101095 (2022)
31. H. Cui et al., SmartMat **3**(4), 565 (2022)
32. P. Mu et al., Energy Fuels **35**(3), 1966 (2021)
33. X. Liu et al., J. Energy Chem. **61**, 104 (2021)
34. H. Shahali et al., Energy Storage Mater., 103130 (2023)
35. T.A. Bendikov et al., J. Phys. Chem. B **106**(11), 2989 (2002)
36. M.M. Gross, A. Manthiram, ACS Appl. Mater. Interfaces **10**(13), 10612 (2018)
37. Y. Zhao et al., Adv. Mater. **32**(32), 2003070 (2020)
38. W. Li et al., Adv. Sci. **7**(23), 2000761 (2020)
39. H. Zhang et al., ACS Nano **16**(5), 7344 (2021)
40. M. Cui et al., ACS Appl. Mater. Interfaces **13**(46), 54981 (2021)
41. A. Amiri et al., ACS Nano **17**(2), 1217 (2022)
42. Z. Xu et al., Chem. Commun. **58**(58), 8145 (2022)
43. W. Wu et al., Energy Environ. Sci. **16**(10), 4326 (2023)
44. P. Cai et al., Adv. Energy Mater. **13**(28), 2301279 (2023)
45. Y. Guo et al., Small, 2207133 (2023)
46. M. Yang et al., Angew. Chem. **134**(42), e202212666 (2022)
47. W. Zhang et al., Adv. Func. Mater. **33**(11), 2210899 (2023)
48. J. Phillips et al., ECS Trans. **16**(16), 11 (2009)
49. Y. Zhang et al., Chem. Eng. J. **451**, 138915 (2023)
50. S.A. Mehta et al., J. Appl. Electrochem.Electrochem. **47**, 167 (2017)
51. J. Scharf et al., Adv. Energy Mater. **11**(33), 2101327 (2021)
52. D. Liu et al., Nano Energy **101**, 107474 (2022)
53. G. Chang et al., Chem. Eng. J. **457**, 141083 (2023)

Chapter 2
Fundamentals of Battery Chemistry

Guided by these achievements, the unique strengths and weaknesses of Zn–S battery are summarized in Fig. 2.1. Here is a brief overview: (a) The primary advantage of sulfur-based battery lies in their compatible potential in water, serving as both cathode and anode when paired with suitable counter electrodes; (b) sulfur demonstrates high capacity through a two-electron transfer reaction, with some cases reaching up to 3044 mAh/g_S due to additional redox contributions; (c) Zn–S battery benefits from enhanced ionic conductivity and distinctive solubility of polysulfides due to water-based solvents, leading to superior kinetics; (d) sulfur-based battery offers excellent safety and nontoxicity.

2.1 Electrochemical Principles

Electrochemical power sources denoted as EPSs serve as devices capable of either storing or transforming chemical energy into electrical energy through electrochemical mechanisms. Their functionality relies on the interchange of electrons between two redox couples, a consequence of simultaneous and involuntary reduction and oxidation processes. According to the Nernst [2] equation, the battery, when charged, exists in a thermodynamically non-equilibrium condition, holding an abundance of free enthalpy, commonly referred to as Gibbs free energy—essentially, chemically stored energy. This surplus plays a pivotal role in propelling the chemical redox reaction toward equilibrium. The inherent spontaneity of these reactions results in an overall diminish in free energy, facilitating the conversion of chemical energy into electrical energy. For this conversion to occur, the two redox couples must be physically segregated into two electrode compartments, often referred to as half cells. Nevertheless, external connectivity is maintained through an electrical circuit, typically constructed with a metal conductor. Internally, an electrolyte, functioning as an ion conductor, orchestrates the movement of ions between the half

© The Author(s), under exclusive license to Springer Nature Switzerland AG 2024   11
A. Amiri et al., *The Zinc–Sulfur Battery*,
SpringerBriefs in Applied Sciences and Technology,
https://doi.org/10.1007/978-3-031-71491-7_2

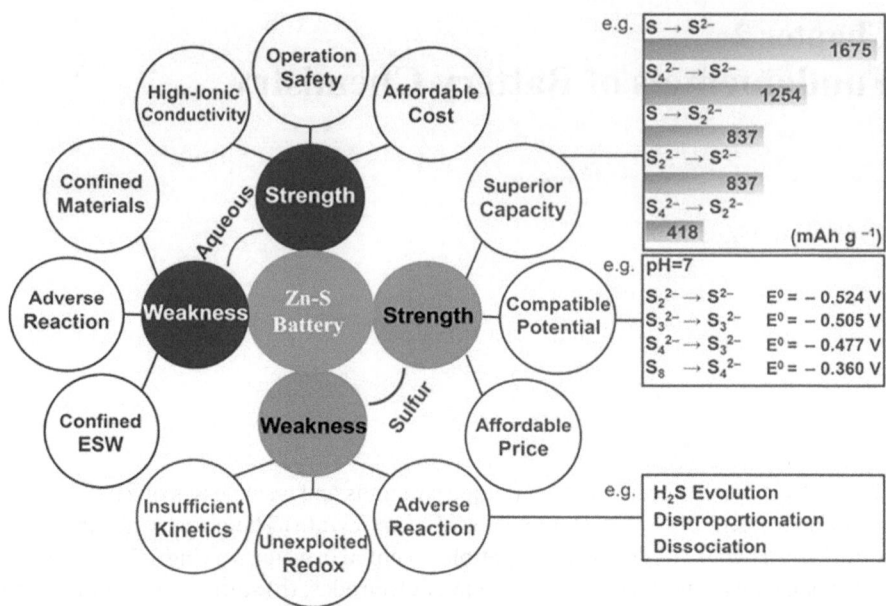

Fig. 2.1 Summary of strengths and weaknesses of Zn–S batteries. Reproduced with permission from [1]

cells, thereby upholding electroneutrality. In essence, electrochemical power sources adeptly leverage the involuntary redox reactions between two couples to generate electrical energy. The arrangement of the two electrodes and the intervening electrolyte constitutes a galvanic cell. The resulting current, denoted as I, is directed toward an external load at a cell voltage E, a value dictated by the specific chemistry of the redox couples engaged. Consequently, this configuration yields the supply of electrical power, symbolized as P, by the cell to a designated end user. In essence, the galvanic cell, with its prescribed electrode and electrolyte arrangement, acts as a source of electrical energy, delivering power to meet the demands of the external load or application [3].

Irrespective of the system selected, the storage of energy through electrochemical means predominantly hinges on interfacial reactions and transport processes, pivotal in determining overall performance. Essentially, all electrochemical systems feature at least two chemically distinct but conducting phases. The boundary between these phases triggers the rearrangement of free charge carriers due to the distinct properties of each phase, leading to the electrification of the interface and the creation of local electric fields. These generated space charges and potential gradients exert significant influence on reactions occurring at or crossing phase boundaries. In electrochemical systems incorporating a solid electrode and liquid electrolyte, the charged area at the interface is termed the electrochemical double layer. The concept of the electrochemical double layer took shape in the twentieth century, originating from Helmholtz's investigations into charge distributions in conductive materials [4]. The term "electrical double layer" was coined to describe the formation of two oppositely charged layers at the surface. Chapman's groundbreaking work in 1913 laid the foundation for the initial double-layer model at the metal–electrolyte interface, incorporating insights from studies on the electrocapillarity of liquids [5–7]. Stern further refined the concept by delineating a fixed and diffusive part of the charged layer in the liquid phase [8]. This historical development illustrates the electrochemical double layer as an inner and outer Helmholtz layer, representing specifically and nonspecifically adsorbed ions, along with a diffuse or Gouy–Chapman layer in the electrolyte.

The initiation of the overall process is set in motion by an autonomous redox reaction, as broadly delineated in Eqs. 2.1–2.3. This reaction involves the interplay of the two redox couples positioned at the respective electrodes. Due to the physical segregation of the half cells, where reduction and oxidation unfold (cathode and anode, respectively), the direct exchange of n electrons is obstructed by the absence of contact between the reactants. Consequently, the necessity arises for the forced passage of these electrons through the external electrical circuit. Essentially, the isolation of the half cells mandates an indirect pathway for electron transmission, which is facilitated by the external electrical circuit, ensuring the seamless continuation of the electrochemical process [3].

$$\text{Cathode: } M^{x+} + ne^- \rightarrow M^{(x-n)+} \tag{2.1}$$

$$\text{Anode: } A^{y-} \rightarrow A^{(y-n)-} + ne^- \tag{2.2}$$

$$\text{Total: } M^{x+} + A^{y-} \rightarrow M^{(x-n)+} + A^{(y-n)-} \tag{2.3}$$

2.2 Components of a Battery

The cell is primarily constituted by two essential components: the electrodes and the electrolyte. Functioning as the dynamic elements of the cell, the electrodes serve as electrical conductors, establishing external connectivity through the electrical circuit and internal linkage via the electrolyte. The pivotal charge transfer processes, involving distinct electron exchanges at the two half cells, occur precisely at the interfaces connecting the electrodes with the electrolyte. There is also a separator that hinders electronic conduction while allowing for ionic conduction when combined with the electrolyte to equalize charge imbalances and complete the electrical circuit [3, 9].

Oxidation occurs in one half-cell, elevating the oxidation state of the involved ionic species and transferring electrons to the corresponding electrode, known as the anode. In the absence of an externally applied electric potential during the discharge process, this electrode becomes negatively charged, representing the electrode with a negative electric potential, also referred to as negative mass. Simultaneously, in the other half-cell, reduction takes place, reducing the oxidation state of the participating ionic species and depleting the electron content in the respective electrode, identified as the cathode. During discharge, this electrode acquires a positive charge, serving as the electrode with a positive electric potential, also termed positive mass. When the battery undergoes charging, an externally applied electric potential reverses ionic migration in the electrolyte, consequently altering the locations of oxidation and reduction, which correspond to the anode and cathode, respectively. Despite these changes, the negative and positive mass characteristics remain unchanged [10].

2.3   Basic Reactions in Zinc–Sulfur Batteries

A comprehensive understanding of the alternations in electrochemical reactions occurring within a Zn–S battery can be achieved through a meticulous examination of classical half-reactions [11]. By delving into the intricacies of these fundamental half-reactions, one can effectively elucidate the dynamic interplay of chemical species and electron transfers, shedding light on the nuanced mechanisms that govern the electrochemical behavior of the Zn–S system. This in-depth analysis not only facilitates the identification of key reaction pathways but also provides valuable insights into optimizing and enhancing the performance of Zn–S batteries for various applications encompassing ESSs:

$$\text{Cathode: } S_{(s)} + 2e^- \rightarrow S^{2-} \quad E^o = -0.58 \text{ V (vs. SHE)} \tag{2.4}$$

$$\text{Anode: } ZnS_{(s)} + 2e^- \rightarrow Zn_{(s)} + S^{2-} \tag{2.5}$$

$$\text{Total: } Zn_{(s)} + S_{(s)} \rightarrow ZnS_{(s)} \quad U^o = 1.04 \text{ V} \tag{2.6}$$

Recent findings from investigations [12–17] have revealed that the second acid dissociation constant of H_2S in aqueous solutions, as outlined in Eq. 2.7, is identified to be less than the widely recognized value, measuring approximately 10^{-17}.

$$HS^- + OH^- \rightarrow S^{2-} + H_2O \quad pK_2 = 17.3 \tag{2.7}$$

These findings imply that the influence of S^{2-} remains virtually inconsequential over a wide range of acidity and sulfide ions concentration, as depicted in the Pourbaix diagram (Fig. 2.2). The Pourbaix diagram provides insight into the complex thermodynamic evolution, a process governed by both the pH condition and the initial chemical reactivity of sulfur-based compounds. Notably, stable states of sulfur-based compounds, encompassing S^{2-}, HS^-, $H_2S_{(aq)}$, SO_4^{2-}, and HSO^{4-}, emerge when submerged in water with an initial chemical reactivity of 1 M sulfur

Fig. 2.2 Pourbaix diagram represents the stability regions of different phases of sulfur under different conditions in an aqueous environment. Reproduced with permission from [1]

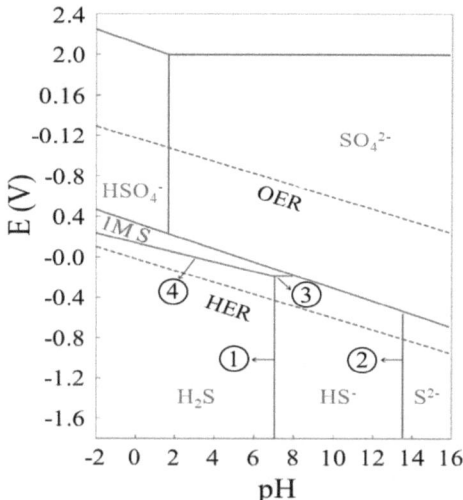

and at a temperature of 25 °C. In this book, the successive phases are delineated as follows: Evolution 1 encompasses the shift from HS^- to H_2S occurring at a pH-neutral condition; Evolution 2 involves the transformation from HS^- to S^{2-} at a pH of 13.90; Evolution 3 corresponds to the transition from dissolved sulfur to HS^- under pH, condition surpassing 7; and Evolution 4 signifies the conversion from dissolved sulfur to H_2S in conditions characterized by moderate acidity. As a result, the pH conditions play a pivotal role in dictating the trajectory of these transformations, shaping the progressions of redox reactions across diverse levels of reactivity in sulfur-based compounds. It is advisable to sustain a relatively high pH level to inhibit the release of H_2S [1].

Moreover, emphasizing the meticulous control and regulation of the initial chemical reactivity (or concentration) of polysulfides is paramount due to its profound impact on the chemical balance [1, 18]. The significance of managing this aspect becomes evident within an aqueous solution, where the optimal concentration spectrum for polysulfide spans from 2 to 25 M. In the electrochemical reduction process, the higher concentration of polysulfide within this range promotes the efficient generation of S_x^{2-}/S^{2-} species. This heightened concentration facilitates the desired electrochemical reactions. On the contrary, lower concentrations play a crucial role in encouraging the formation of S_x^{2-}/HS^-, influencing the electrochemical reduction toward the desired pathways [19].

Furthermore, it is imperative to acknowledge that, in addition to the release of H_2S, the concurrent thermodynamic phenomena of disproportionation and dissociation, articulated through Eqs. 2.8 and 2.9, exert a significant impact on diminishing the operational longevity of sulfur-aqueous-based batteries (SABs). Thus, the natural progression of thermodynamics within SABs results in an inherent reduction in the accessibility of active components. Fortunately, the adjustment of pH levels and chemical reactivity provides a way to steer redox reactions on a favorable course, reducing the impact of adverse reactions to a certain extent. Nevertheless, it is crucial to note that overly high pH values and polysulfide concentrations may introduce additional complications, encompassing constraints in the selection of appropriate materials for the battery components and an augmentation in the viscosity of the electrolyte.

$$S_{dissolved} + OH^- \rightarrow \frac{1}{4}S_2O_3^{2-} + \frac{1}{2}HS^- + \frac{1}{4}H_2O \tag{2.8}$$

$$S_6^{2-} \rightarrow 2S_3^- \tag{2.9}$$

$$S_6^{2-} \rightarrow S_5^- + S_7^- \tag{2.10}$$

The electrochemical transformations within Zn–S systems, particularly those utilizing aqueous electrolytes, exhibit a notable simplicity in their inherent nature. In an investigation performed by Yang et al. [20], ex situ X-ray Diffraction (XRD) was employed to meticulously scrutinize the cathode, shedding light on the transformation mechanism of the sulfur cathode within the electrolytic environment, as illustrated in Fig. 2.3a. Throughout the discharge process, the main sulfur peak's intensity gradually decreased, giving way to the emergence of a signal associated with zinc sulfide (ZnS). As the discharge continued, the prominence of the ZnS peak increased, and the sulfur signal was completely eradicated. The return of the sulfur peak as the electrode was brought back to 1.6 V signified a reversible transformation between sulfur and ZnS. Further validation of the solid-to-solid transformation occurring in the sulfur cathode throughout the electrochemical procedure was achieved through the development of Raman spectra, as illustrated in Fig. 2.3b, c. Li and collaborators [21], alongside other studies, substantiated this occurrence using ex situ XRD, Raman spectroscopy, Transmission Electron Microscopy (TEM) examination, as portrayed in Fig. 2.3d, and X-ray Photoelectron Spectroscopy (XPS) analysis. XPS analysis revealed the manifestation of a zinc peak through the discharge process, signifying the electrochemical incorporation and subsequent removal of zinc from the cathode material. Furthermore, the discharge phase revealed the presence of the S^{2-} signal, whereas the charging phase exhibited the emergence of the SO_4^{2-} signal. This indicates the potential oxidation of certain surface sulfur, leading to the formation of SO_4^{2-}, as depicted in Fig. 2.3c. Through in-depth and in situ observations, it was identified that the cathode's interatomic spacing at 0.3145 closely aligned with the (111) lattice plane of ZnS (Fig. 2.3e). This underscores that the aqueous Zn–S solution battery, featuring an elemental sulfur cathode, metallic zinc anode, and a mild electrolyte, fundamentally participates in a solid-state redox reaction involving two electrons. The prominent sulfide constituents within the aqueous solution encompass HS^- and OH^-. The interaction between the aqueous electrolyte and ZnS is more precisely explained through the following equations [22]:

$$\text{Cathode:} \ S_{(s)} + H_2O + 2e^- \rightarrow HS^- + OH^- \quad E^\circ = -0.48 \, \text{V (vs. SHE)} \tag{2.11}$$

$$\text{Anode: } ZnS_{(s)} + H_2O + 2e^- \rightarrow Zn_{(s)} + HS^- + OH^- \quad E^o = -1.52\,V \text{ (vs. SHE)} \tag{2.12}$$

$$\text{Total: } Zn_{(s)} + S_{(s)} \rightarrow ZnS_{(s)} \quad U^o = 1.04\,V \tag{2.13}$$

Therefore, despite the inclusion of water in Eqs. 2.11 and 2.12, it remains inactive throughout the recharging process. As indicated by the discharge mechanisms outlined in Eqs. 2.14 and 2.15, the electrochemical performance of Zn–S demonstrates a faradaic capacity measuring 550 Ah/kg, along with a theoretical energy density of 572 Wh/kg [23].

$$\text{Cathode: } S + Zn^{2+} + 2e^- \rightarrow ZnS \tag{2.14}$$

$$\text{Anode: } Zn - 2e^- \rightarrow Zn^{2+} \tag{2.15}$$

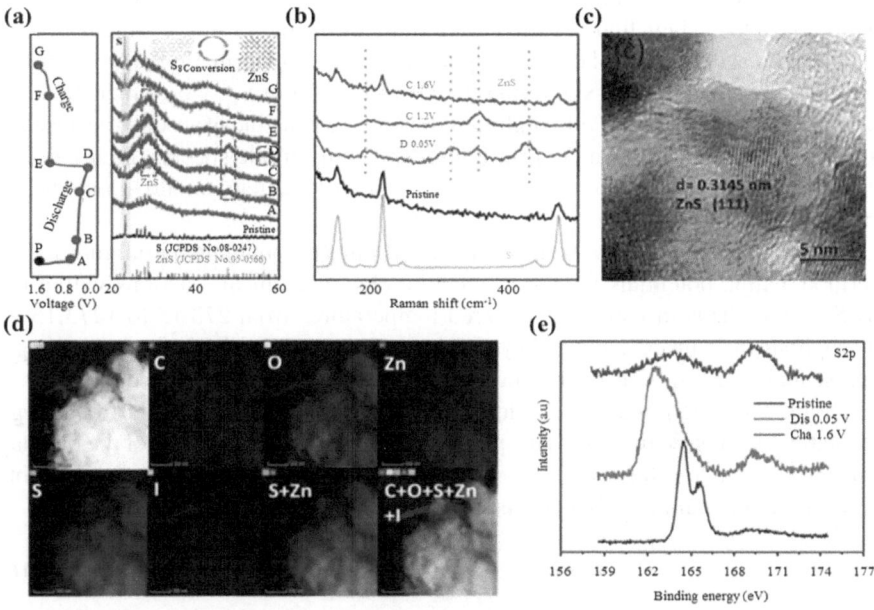

Fig. 2.3 a Recorded XRD spectra of the sulfur cathode under certain potentials. The left section portrays unique voltage profiles concerning the Zn–S cell. Designated points on these profiles signify the selected states for subsequent ex situ analyses. **b** Raman spectra for both sulfur and the hollow carbon spheres (HCS)/S-53.7 electrodes under diverse charging and discharging states [20]. **c** TEM images and **d** TEM mapping displaying the complete discharge state for S@CNTs-50. **e** XPS analyses of S 2p spectra in their original, completely depleted, and recharged conditions [21]. Reproduced with permission from [20, 21]

In contrast to the theoretical energy capacities of conventional aqueous cell systems, which encompass Pb-acid (170 Wh/kg), Ni–Cd (217 Wh/kg), alkaline Zn–MnO$_2$ (336 Wh/kg), and zinc-silver (447 Wh/kg), the Zn–S electrochemical storage system stands out with a significant increase in energy density [24]. At a temperature of 298.15 K, this system displays a distinctive thermodynamic inclination toward the creation of ZnS product ($\Delta G^o = -201.3$ kJ/mol) as opposed to the formation of Zn(OH)$_4^{2-}$ or Zn(OH)$_2$ ($\Delta G^o = -156.3$ kJ/mol). This behavior is in stark contrast to aluminum–sulfur (Al–S) systems. Consequently, it is expected that the Zn–S reaction will refrain from utilizing hydroxide through the discharging process, resulting in an improvement in the effectiveness of utilizing electroactive materials. Furthermore, considering the aqueous voltage loss within Al–S systems, the energy density of an Al–S cell, measured at 507 Wh/kg, falls below that of the Zn–S electrochemical storage, offering an alternative. The advantageous features of the Zn–S system suggest that it may be a more enticing alternative than the Al–S system within the realm of electrochemical power generation.

Despite the documentation of diverse mechanisms during the discharging of Zn–S batteries, they consistently correspond with the observations made by Gross and Manthiram [19]. Their findings unveil a unique discharge profile distinguished by two plateaus, with the initial one corresponding to the transformation of S^{2-} to S$_4^{2-}$, and the subsequent plateau signifying the alternation from S$_4^{2-}$ to elemental sulfur (S^0). Moreover, it is essential to highlight that even when zinc is absent, the accessibility of thermodynamic data concerning diverse polysulfide species, their equilibrium, or relevant half-reactions is constrained to particular temperature intervals and circumstances [25–35].

In their study, Bendikov et al. [30] carried out computations pertaining to the thermodynamic potentials linked to potential electrochemical reactions within the Zn–S system. This investigation covered temperatures from 273.15 to 1273.15 K. Throughout their investigation, they determined the standardized reaction voltage, represented as E^o in volts, corresponding to a specific temperature denoted as T in Kelvin. The determination entailed computing the standard reaction free energy alternation, expressed in J/mol. This procedure included the utilization of the Faraday constant, set at a value of $F = 9.649 \times 10^4$ C/mol, along with the variable representing the number of electrons, denoted as n:

$$E^o(T) = -\Delta G^o(T)_{\text{reaction}}/nF \qquad (2.16)$$

Within their investigation, the value of $\Delta G^o(T)_{\text{reaction}}$ was ascertained through the subtraction of $\sum \Delta G^o(T)_{\text{products}}$ from $\sum \Delta G^o(T)_{\text{reactants}}$. This computational process was performed individually for each involved species engaged in the reaction. This computation encompassed the distinction between the alternation in standard enthalpy of formation, denoted as ΔH^o (expressed in J/mol), and $T\Delta S^o(T)$, utilizing the change in standard entropy (expressed in J/mol K). The assessment of the alternation in standard enthalpy and standard entropy, denoted as S^o (expressed in J/mol K), entailed integrating a consistent heat capacity, C_p (expressed in J/mol K), as either $C_p dT$ or $C_p dT/t$. This integration was carried out across the temperature range, commencing from 298.15 K [30].

Their observations unveiled that sulfur behaves as an insulator under room temperature conditions. Although it possesses a considerable intrinsic storage capacity of two Faraday/mol ($FW(S) = 32.07$ g/mol), employing it as a cathode presents drawbacks. Nevertheless, establishing a proficient interfacial connection at or around room temperature could be facilitated by dissolving of elemental sulfur into diverse soluble aqueous polysulfide compounds. Their study additionally entailed forecasting discharge reactions that align with thermodynamic preferences. Intricate speciation and equilibrium characterize aqueous alkaline solutions containing sulfur and sulfide salts, denoted as $M_X S_Y$. This involves entities such as $M^{2y/x}$, H_2S, HS^-, S^{2-}, S_2^{2-}, S_3^{2-}, S_4^{2-}, S_5^{2-}, H_2O, H^+, and OH^-. These entities participate in electrochemical processes with zinc, resulting in the generation of diverse products. This book has identified potential electrochemical reactions that may take place in this realm. The ascertained thermodynamic potentials for 29 diverse electrochemical reactions in aqueous environments at a temperature of 273.15 K are illustrated in Table 2.1. The reactions fall into four primary groups concerning zinc discharge. Categories "i" through "iii" yield ZnS discharge products, whereas the fourth group, "iv," leads to the creation of either a zincate or zinc hydroxide product. Within Table 2.1, reactions labeled as "i" entail the oxidation of a polysulfide reactant, resulting in the production of a polysulfide product with a reduced chain length [26, 27, 32–35].

$$Zn_{(s)} + S_{x(aq)}^{2-} \rightarrow ZnS_s + S_{(x-1)(aq)}^{2-} \tag{2.17}$$

Reactions categorized as "ii" encompass the oxidative transformation of a polysulfide reactant, leading to the formation of a hydrosulfide product.

$$(x-1)Zn_{(s)} + S_{x(aq)}^{2-} + H_2O \rightarrow (x-1)ZnS_{(s)} + HS_{(aq)}^- + OH_{(aq)}^- \tag{2.18}$$

Reactions classified under the category "iii" involve the oxidative conversion of a polysulfide reactant, resulting in the production of a sulfide product.

$$(x-1)Zn_{(s)} + S_{x(aq)}^{2-} \rightarrow (x-1)ZnS_{(s)} + S_{(aq)}^{2-} \tag{2.19}$$

Table 2.1 reveals a consistent trend of elevated cell potentials for reactions in the "i–iii" categories in contrast to category "iv." This pattern distinctly indicates a strong energetic preference for the generation of ZnS products over $Zn(OH)_4^{2-}$ or $Zn(OH)_2$ products across various scenarios.

The arrangement of polysulfide entities within aqueous solutions incorporating polysulfides undergoes a discernible impact due to the alkaline nature of the solution [26, 33, 34, 36]. The dominant species is chiefly contingent on the solution's pH, exhibiting distinct patterns across various pH spectrums. For example, within the pH spectrum spanning from 9 to 14, the tetrasulfide variant (S_4^{2-}) assumes a

Table 2.1 Anticipated reactions within the Zn–S system under aqueous alkaline conditions and their corresponding thermodynamic potentials at a temperature of 273.15 K. Reproduced with permission from [30]

	Equation	E^o_{reaction} at 273.15 K
	$Zn_{(s)} + S_{(s)} \rightarrow ZnS_{(s)}$	1.048
i-1	$Zn_{(s)} + S_3^{2-}{}_{(aq)} \rightarrow ZnS_{(s)} + S_2^{2-}{}_{(aq)}$	1.016
i-2	$Zn_{(s)} + S_4^{2-}{}_{(aq)} \rightarrow ZnS_{(s)} + S_3^{2-}{}_{(aq)}$	1.021
i-3	$Zn_{(s)} + S_5^{2-}{}_{(aq)} \rightarrow ZnS_{(s)} + S_4^{2-}{}_{(aq)}$	1.029
ii-1	$Zn_{(s)} + S_2^{2-}{}_{(aq)} + H_2O_{(l)} \rightarrow$ $ZnS_{(s)} + HS^-_{(aq)} + OH^-_{(aq)}$	0.798
ii-2	$2Zn_{(s)} + S_3^{2-}{}_{(aq)} + H_2O_{(l)} \rightarrow$ $2ZnS_{(s)} + HS^-_{(aq)} + OH^-_{(aq)}$	0.907
ii-3	$3Zn_{(s)} + S_4^{2-}{}_{(aq)} + H_2O_{(l)} \rightarrow$ $3ZnS_{(s)} + HS^-_{(aq)} + OH^-_{(aq)}$	0.945
ii-4	$4Zn_{(S)} + S_5^{2-}{}_{(aq)} + H_2O_{(l)} \rightarrow$ $4ZnS_{(s)} + HS^-_{(aq)} + OH^-_{(aq)}$	0.966
iii-1	$Zn_{(s)} + S_2^{2-}{}_{(aq)} \rightarrow ZnS_{(s)} + S^{2-}_{(aq)}$	0.826
iii-2	$2Zn_{(s)} + S_3^{2-}{}_{(aq)} \rightarrow 2ZnS_{(s)} + S^{2-}_{(aq)}$	0.936
iii-3	$3Zn_{(s)} + S_4^{2-}{}_{(aq)} \rightarrow 3ZnS_{(s)} + S^{2-}_{(aq)}$	0.954
iii-4	$4Zn_{(s)} + S_5^{2-}{}_{(aq)} \rightarrow 4ZnS_{(s)} + S^{2-}_{(aq)}$	0.973
iv-1	$Zn_{(s)} + S_{(s)} + 3OH^-_{(aq)} + H_2O_{(l)} \rightarrow$ $Zn(OH)_4^{2-}{}_{(aq)} + HS^-_{(aq)}$	0.731
iv-2	$Zn_{(s)} + S_2^{2-}{}_{(aq)} + 2OH^-_{(aq)} + 2H_2O_{(l)} \rightarrow$ $Zn(OH)_4^{2-}{}_{(aq)} + 2HS^-_{(aq)}$	0.498
iv-3	$Zn_{(s)} + S_3^{2-}{}_{(aq)} + 3OH^-_{(aq)} + H_2O_{(l)} \rightarrow$ $Zn(OH)_4^{2-}{}_{(aq)} + S_2^{2-}{}_{(aq)} + HS^-_{(aq)}$	0.716
iv-4	$Zn_{(s)} + S_4^{2-}{}_{(aq)} + 3OH^-_{(aq)} + H_2O_{(l)} \rightarrow$ $Zn(OH)_4^{2-}{}_{(aq)} + S_3^{2-}{}_{(aq)} + HS^-_{(aq)}$	0.723
iv-5	$Zn_{(s)} + S_5^{2-}{}_{(aq)} + 3OH^-_{(aq)} + H_2O_{(l)} \rightarrow$ $Zn(OH)_4^{2-}{}_{(aq)} + S_4^{2-}{}_{(aq)} + HS^-_{(aq)}$	0.729
iv-6	$2Zn_{(s)} + S_3^{2-}{}_{(aq)} + 5OH^-_{(aq)} + 3H_2O_{(l)} \rightarrow$ $2Zn(OH)_4^{2-}{}_{(aq)} + 3HS^-_{(aq)}$	0.607

(continued)

Table 2.1 (continued)

	Equation	$E^o_{reaction}$ at 273.15 K
iv-7	$3Zn_{(s)} + S^{2-}_{4\ (aq)} + 8OH^-_{(aq)} + 4H_2O_{(l)} \rightarrow$ $3Zn(OH)^{2-}_{4\ (aq)} + 4HS^-_{(aq)}$	0.645
iv-8	$4Zn_{(s)} + S^{2-}_{5\ (aq)} + 11OH^-_{(aq)} + 5H_2O_{(l)} \rightarrow$ $4Zn(OH)^{2-}_{4\ (aq)} + 5HS^-_{(aq)}$	0.666
iv-9	$Zn_{(s)} + S_{(s)} + 4OH^-_{(aq)} \rightarrow Zn(OH)^{2-}_{4\ (aq)} + S^{2-}_{(aq)}$	0.761
iv-10	$Zn_{(s)} + S^{2-}_{2\ (aq)} + 4OH^-_{(aq)} \rightarrow$ $Zn(OH)^{2-}_{4\ (aq)} + 2S^{2-}_{(aq)}$	0.556
iv-11	$Zn_{(s)} + S^{2-}_{3\ (aq)} + 4OH^-_{(aq)} \rightarrow$ $Zn(OH)^{2-}_{4\ (aq)} + S^{2-}_{2\ (aq)} + S^{2-}_{(aq)}$	0.745
iv-12	$Zn_{(s)} + S^{2-}_{4\ (aq)} + 4OH^-_{(aq)} \rightarrow$ $Zn(OH)^{2-}_{4\ (aq)} + S^{2-}_{3\ (aq)} + S^{2-}_{(aq)}$	0.751
iv-13	$Zn_{(s)} + S^{2-}_{5\ (aq)} + 4OH^-_{(aq)} \rightarrow$ $Zn(OH)^{2-}_{4\ (aq)} + S^{2-}_{4\ (aq)} + S^{2-}_{(aq)}$	0.758
iv-14	$2Zn_{(s)} + S^{2-}_{3\ (aq)} + 8OH^-_{(aq)} \rightarrow$ $2Zn(OH)^{2-}_{4\ (aq)} + 3S^{2-}_{(aq)}$	0.651
iv-15	$3Zn_{(s)} + S^{2-}_{4\ (aq)} + 12OH^-_{(aq)} \rightarrow$ $3Zn(OH)^{2-}_{4\ (aq)} + 4S^{2-}_{(aq)}$	0.684
iv-16	$4Zn_{(s)} + S^{2-}_{5\ (aq)} + 16OH^-_{(aq)} \rightarrow$ $4Zn(OH)^{2-}_{4\ (aq)} + 5S^{2-}_{(aq)}$	0.702
iv-17	$4Zn_{(s)} + 2H_2O_{(l)} \rightarrow Zn(OH)_{2(aq)} + H_{2(g)}$	0.414

leading role, whereas, in the conditions approaching neutrality to a slightly alkaline state, S^{2-}_5 emerges as the prevailing variant. In solutions featuring elevated pH levels (approximately 14–16), with a considerable presence of hydroxide ions, S^{2-}_3 takes on the role of a primary variant. Under exceedingly high pH conditions (pH > 16), typically encountered in concentrated and highly reactive hydroxide solutions, the prevailing variant consists of S^{2-}_3 and S^{2-}_2 [37]. The significant reactivity of S^{2-} is observed exclusively under conditions of exceptionally elevated pH, whereas at lower pH levels, sulfur undergoes complete reduction and exists in its hydrolyzed state as HS^-. In light of these revelations and the thermodynamic potentials outlined in Table 2.1, conjectures are devised concerning the preferred reactions within the cell.

A prominent hurdle in a Zn–S battery electrolyte is the initiation of the HER, commonly activated in specific situations, like over-discharge or when the stability of the electrolyte is jeopardized. The outcome of this procedure is the production of hydrogen gas, a situation not only inefficient but also entails safety concerns [38, 39]. The emergence of HER and the growth of dendrite structures in aqueous Zn-based batteries are substantially affected by the composition, pH, and potential window of the electrolyte. Recently published articles delved into methodologies designed to enhance the stability of zinc anodes within such batteries [40, 41]. Gross and Manthiram [19] undertook an extensive examination of the operational mechanisms associated with polysulfides, especially in their groundbreaking research on rechargeable zinc-aqueous polysulfide (ZAPS) batteries. During this examination, they pioneered and elucidated the idea of ZAPS batteries, unraveling their electrochemical traits. Throughout the discharge process, zinc experiences oxidation, resulting in the formation of the zincate ion as per the reaction:

$$Zn + 4OH^- \rightarrow [Zn(OH)_4]^{2-} + 2e^- - 1.22\,V \text{ vs. SHE} \tag{2.20}$$

They pointed out that, on the cathode, the prevailing S_4^{2-} undergoes a transformation to S^{2-} using a simple reaction, especially relevant in scenarios with exceptionally high polysulfide concentrations.

$$S^{2-} + 6e^- \rightarrow 4S^{2-} \tag{2.21}$$

They further asserted that the solubilization of sulfur in aqueous environments, when sulfides are present, leads to the formation of diverse entities, encompassing HS^-, H_2S, S_2^-, and polysulfide species S_X^{2-} ($x = 2 - 5$). They also highlighted that the dispersion of these entities is notably impacted by the acidity level of the solution, as well as the initial amount of sulfide and elemental sulfur. The electrochemical response is ascribed to the existence of sulfide and polysulfide compounds, as outlined in.

$$S^{2-} + 2e^- \rightarrow 2S^{2-} \quad 0.52\,V \text{ vs. SHE} \tag{2.22}$$

$$3S^{2-} + 2e^- \rightarrow 4S^{2-} \quad 0.48\,V \text{ vs. SHE} \tag{2.23}$$

$$2S^{2-} + 2e^- \rightarrow 3S^{2-} \quad 0.50\,V \text{ vs. SHE} \tag{2.24}$$

References

1. J. Liu et al., J. Am. Chem. Soc. **143**(38), 15475 (2021)
2. W. Nernst, Ber. Dtsch. Chem. Ges. **30**(2), 1547 (1897)
3. F. Nobili, R. Marassi, Batteries: present and future energy storage challenges **2**, 13 (2020)
4. H.V. Helmholtz, Annalen der Physik **165**(7), 353 (1853)

5. G. Lippmann, Beziehungen zwischen den capillaren und elekrischen Erscheinungen (1873)
6. M. Gouy, J. Phys. Theor. Appl. **9**(1), 457 (1910)
7. D.L. Chapman, London Edinburgh Dublin Philosoph. Mag. J. Sci. **25**(148), 475 (1913)
8. O. Stern, Z. Elektrochem. **30**(508), 1014 (1924)
9. M. Zschornak et al., Phys. Sci. Rev. **3**(11), 20170111 (2018)
10. F. Meutzner et al., Electrochem. Storage Mater. From Crystallogr. Manuf. Technol. (2018)
11. S. Licht, J. Electrochem. Soc. **135**(12), 2971 (1988)
12. D. Peramunage et al., Anal. Chem. **66**(3), 378 (1994)
13. S. Licht et al., J. Electroanal. Chem. Interfacial Electrochem. **318**(1–2), 111 (1991)
14. S. Licht et al., Anal. Chem. **62**(13), 1356 (1990)
15. S. Licht, J. Manassen, J. Electrochem. Soc. **134**(4), 918 (1987)
16. B. Meyer et al., Inorg. Chem. **22**(16), 2345 (1983)
17. W. Giggenbach, Inorg. Chem. **10**(7), 1333 (1971)
18. S. Licht, J. Davis, J. Phys. Chem. B **101**(14), 2540 (1997)
19. M.M. Gross, A. Manthiram, ACS Appl. Mater. Interfaces **10**(13), 10612 (2018)
20. M. Yang et al., Angew. Chem. **134**(42), e202212666 (2022)
21. W. Li et al., Adv. Sci. **7**(23), 2000761 (2020)
22. X. Zheng et al., Sci. China Mater. **65**(6), 1463 (2022)
23. D. Patel, A.K. Sharma, Energy Fuels **37**(15), 10897 (2023)
24. W. Shi et al., Chemsuschem **14**(7), 1634 (2021)
25. V. Sharivker et al., Electrochim. Acta **41**(15), 2381 (1996)
26. S. Licht, J. Electrochem. Soc. **134**(9), 2137 (1987)
27. S. Licht et al., Inorg. Chem. **25**(15), 2486 (1986)
28. A.J. Bard et al., *Copper, Silver, and Gold. In Standard Potentials in Aqueous Solution* (Routledge, 2017), p. 287
29. P.M. Lessner et al., J. Electrochem. Soc. **140**(7), 1847 (1993)
30. T.A. Bendikov et al., J. Phys. Chem. B **106**(11), 2989 (2002)
31. D.A. Probst, G. Henderson, J. Chem. Educ. **73**(10), 962 (1996)
32. A. Teder, The equilibrium between elementary sulfur and aqueous polysulfide solutions. Svenska Träforskningsintitutet (1971)
33. W. Giggenbach, Inorg. Chem. **11**(6), 1201 (1972)
34. P. Lessner et al., J. Electrochem. Soc. **133**(12), 2517 (1986)
35. G. Le Guillanton et al., J. Electrochem. Soc. **143**(10), L223 (1996)
36. S. Licht, Nature **330**(6144), 148 (1987)
37. S. Licht, Anal. Chem. **57**(2), 514 (1985)
38. W. Zhang et al., Adv. Func. Mater. **33**(11), 2210899 (2023)
39. Y. Zhao et al., Adv. Mater. **32**(32), 2003070 (2020)
40. X. Zhang et al., InfoMat **4**(7), e12306 (2022)
41. C. Xie et al., Carbon Energy **2**(4), 540 (2020)

Chapter 3
Zinc–Sulfur Battery Design and Construction

3.1 Cell Components and Materials

As detailed in Sect. 2.2 of this book, the batteries' fundamental architecture is composed of two indispensable elements: the electrolyte and the electrodes, which are presented in this chapter.

3.2 Electrolytes and Separators

In Zn–S batteries, the electrolyte stands out as a key element, chiefly tasked with enabling the transfer of ions between electrodes, setting up an electrochemically stable potential window (ESPW), and improving the overall electrochemical efficiency. Consequently, opting for a well-suited electrolyte has the potential to enhance the reversibility during zinc plating and stripping, regulate reaction mechanisms effectively, and augment ionic conductivity. Fundamentally, the precision and efficiency of Zn–S batteries hinge significantly on the careful selection of an appropriate electrolyte [1]. Moreover, another integral component in the system is the separator, which impedes electronic conduction while enabling ionic conduction in tandem with the electrolyte. By strategically allowing the passage of ions and preventing the direct flow of electrons, the separator acts as a pivotal element, maintaining the overall functionality, balance of electrochemical processes, and the completion of the electrical circuit in the battery system [2].

Broadly speaking, the electrolytes applicable in Zn–S batteries can be classified into three principal categories, as illustrated in Fig. 3.1. These classifications include aqueous electrolytes, both with and without redox mediators such as Thiourea (TU), Iodinated Thiourea (TUI), di-iodine (I_2), among others, hydrogel electrolytes, and organic/non-polar electrolytes exemplified by Deep Eutectic Solvents (DES), incorporating affordable constituents such as urea and choline chloride (ChCl).

© The Author(s), under exclusive license to Springer Nature Switzerland AG 2024   25
A. Amiri et al., *The Zinc–Sulfur Battery*,
SpringerBriefs in Applied Sciences and Technology,
https://doi.org/10.1007/978-3-031-71491-7_3

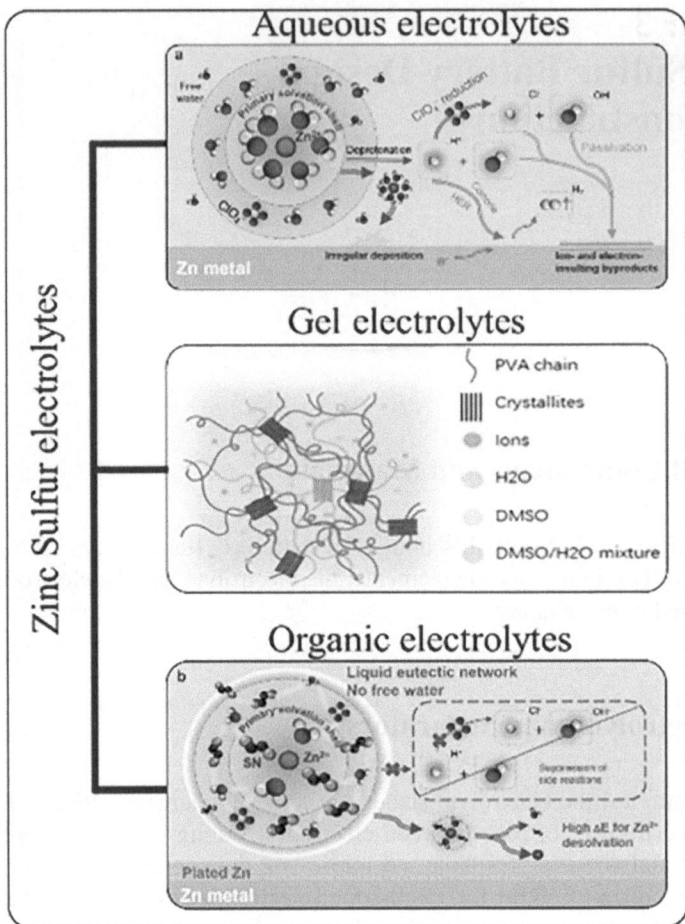

Fig. 3.1 Principal classifications of electrolytes applicable in Zn–S batteries (some graphical elements are reproduced with permission from [5–8])

Aqueous electrolytes with zinc are widely employed in Zn–S batteries, known for their stable features and electrode compatibility. However, these electrolytes present certain drawbacks when applied to the practical functioning of these batteries. Prominent drawbacks encompass substantial polarization resulting from the sluggish reaction of ZnS to S, as outlined in Eq. 2.13, diminished CE, and a rapid deterioration in capacity over time. The origins of these challenges stem from the cathode dissolving in the electrolyte and the emergence of dendritic structures on the zinc anode, phenomena extensively documented in the literature [3]. Beyond conventional aqueous electrolytes, the development of ductile and wearable electronic instruments has highlighted the demand for energy storage devices that exhibit high performance, are lightweight, and are exceptionally thin to facilitate bending. The conventional

Zn–S batteries do not effectively fulfill these requirements. Therefore, scientists are delving into pioneering designs for Zn–S batteries, aiming to incorporate configurations that are lightweight, thin, and integrated while ensuring ductility and stretchability [4]. These atypical designs frequently utilize solid-state (gel) electrolytes and/ or incorporate functional additives into the electrolyte composition, with the objective of addressing the constraints observed in conventional Zn–S batteries.

The advancement of Zn–S battery electrolytes is in its initial phase, requiring comprehensive, thorough, and in-depth research to achieve a full understanding of how electrolytes influence the electrochemical efficiency of Zn–S batteries. This section of the book extensively explores the electrochemical fundamentals governing Zn–S batteries while presenting a comprehensive survey of recent breakthroughs in electrolyte varieties, covering liquid, gel, and adaptable hydrogel formulations. Moreover, the enduring obstacles and future possibilities linked to Zn–S battery electrolytes are addressed, and invaluable perspectives are provided to propel the advancements of Zn–S technology.

A primary research objective now revolves around the quest for Zn–S battery electrolytes capable of synergizing the advantages inherent in a significant sulfur capacity (measuring at 1675 mAh/g) with the affordability (approximately $2/kg) and safety features associated with zinc metal, especially aligning with the safety and non-flammability attributes commonly found in aqueous electrolytes [9]. In this vein, the prospects of creating aqueous Zn–S batteries that possess heightened capacity, safety, and cost-effectiveness carry considerable potential. However, notwithstanding the merits of such batteries, numerous significant hurdles impede their extensive adoption, demanding meticulous consideration and resolution. Concerning zinc-polysulfide batteries (ZPBs), a hurdle arises in the interaction between zinc and polysulfide, as delineated in Sect. 2.3. Resulting from this interaction, a ZnS passive stratum forms on the surface of the zinc anode. Functioning as an impediment, this stratum obstructs subsequent discharge processes and compromises the potential for reversibility [10]. Another hurdle is entwined with the intricate electrochemical processes of sulfur in aqueous electrolytes, including the kinetic and thermodynamic transformations explained in Eqs. 2.17–2.19. Tackling these hurdles highlights the significance of developing a reliable zinc electrode and a well-matched electrolyte capable of synergy with sulfur-based cathodes. Consequently, a thorough evaluation of the discharge–charge mechanisms in these sophisticated Zn–S batteries becomes imperative to attain a comprehensive and all-encompassing understanding of their functionality. In general, this chapter delves into the advancement of zinc-ion electrolytes in each of their classifications, scrutinizing both merits and drawbacks, as depicted in Fig. 3.2.

3.2.1 Aqueous Electrolytes

Aqueous Zn–S batteries, alongside their counterpart ZAPS batteries, are both classified with the broader classification of aqueous Zn–S batteries [12]. Figure 3.3a,

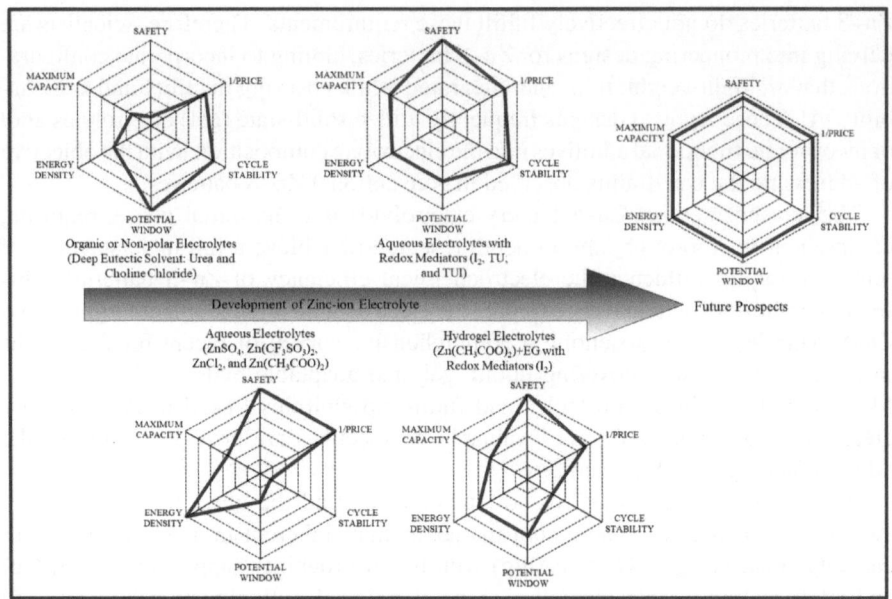

Fig. 3.2 The evolution of electrolytes for Zn–S batteries and their outlook. Reproduced with permission from [11]

b provide schematic depictions delineating the distinctive constituents and electro-chemical principles inherent in these battery variants. Concerning the illustration in Fig. 3.3a, the aqueous Zn–S battery is commonly structured with a zinc anode, a cathode incorporating sulfur, and an aqueous electrolyte. In the discharge stage, reaction as described in Eq. 2.4 unfolds at the cathode electrode surface, whereas at the anode electrode surface, reaction articulated in Eq. 2.5 transpires [13, 14].

Xu and researchers [15] examined the application of concentrated $Zn(OTF)_2$ as an electrolytic solution in Zn–S batteries. In contrast to conventional electrolytes, $Zn(OTF)_2$ displayed superior ionic conductivity, playing a pivotal role in facili-tating the kinetics of Zn^{2+} transportation. This initiative yielded various positive outcomes, such as heightened overpotential for zinc nucleation and a broadened ESPW. Furthermore, the electrolyte effectively inhibited both hydrogen evolution and corrosion at the zinc anode, consequently resulting in improved overall cycling performance of the battery. Noteworthy is the fact that the capacity retention of $Zn(OTF)_2$ outperformed that of $Zn(AC)_2$ and $ZnSO_4$ electrolytes. This superiority can be ascribed to the larger anionic diameter of CF_3SO^{3-} anion present in $Zn(OTF)_2$, which resulted in a reduction of water molecules in Zn^{2+}'s solvation shell, thereby mitigating its solvation impact [16]. Additionally, the ionic conductivity of $Zn(OTF)_2$ was elevated at 60.9 mS/cm, promoting the transport kinetics of Zn^{2+}. However, the persistent challenge of undesirable side reactions occurring between the sulfur cathode and aqueous electrolyte in Zn–S batteries remains, ultimately resulting in less-than-optimal cycling stability.

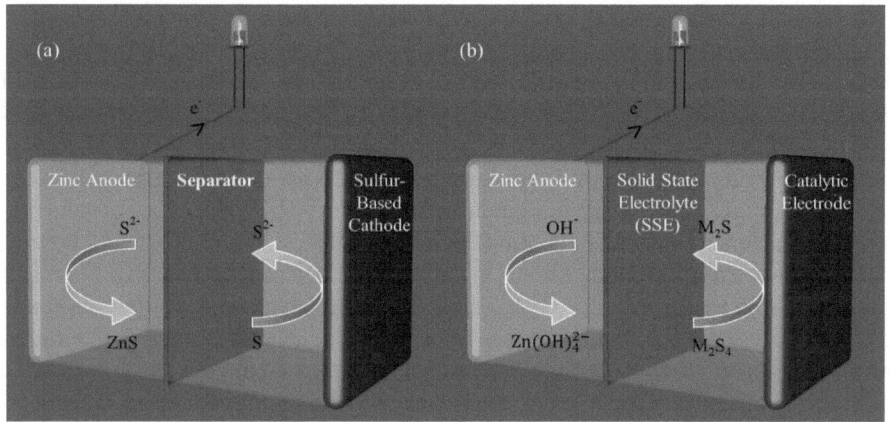

Fig. 3.3 Schematics representing two classifications of Zn-based batteries. **a** Aqueous Zn–S battery, **b** ZAPS battery incorporating a Solid-State Electrolyte (SSE). The "M" in **b** denotes metallic elements like Na and Li (conceived by our research group). Reproduced with permission from [11]

Yang et al. [17] presented a methodology known as "cocktail optimization" for the synthesis of electrolytic solutions. This methodology integrates tetraglyme (G_4) and water as co-solvents, with the incorporation of I_2 as an additive. This novel approach utilized the merits of co-solvents that are miscible with water, along with polar catalysts, to successfully mitigate side reactions and corrosion triggered by water exposure. Furthermore, the inclusion of G_4 played a significant role in preventing the growth of dendrite structures and promoting the efficiency of sulfur conversion kinetics. Particularly noteworthy is G_4's pronounced attraction to the sulfur cathode, significantly ameliorating the wettability of the electrode. This amelioration accelerated the movement of zinc ions within the battery. Another advantage could be seen in this amelioration is the effective mitigation of by-products like SO_4^{2-} throughout the sulfur transformation, attributed to the reduction in infiltration of water.

Guo et al. [18] introduced a methodology for designing a hybrid electrolyte in the Zn–S system, yielding improved electrochemical performance, encompassing a capacity of 1435 mAh/g and an energy density reaching 730 Wh/kg. In implementing this methodology, an aqueous system derived from $Zn(SF_3SO_3)_2$ was adopted, incorporating ethylene glycol (EG), chemically denoted as $(CH_2OH)_2$, as a co-solvent. Additionally, the inclusion of zinc iodide (ZnI_2) served as an additive, coupled with a nonporous carbon-supported sulfur cathode (S@NPC). The introduction of EG into the electrolyte system bolstered hydrogen bonding, restrained the water's proton activity, and expedited the transformation of sulfur. Consequently, there was a broadening of the ESPW, accompanied by significant suppression of sulfate and sulfide anion formation. Additionally, the magnitude of both OER and HER in water was markedly restrained. This led to better capacity retention, outperforming that of pure aqueous electrolytes within the Zn/S setup. Furthermore, microscopic, and spectroscopic investigations disclosed that in a hybrid electrolyte with a 50% volume ratio (HE-50), where EG constituted half of the solvent volume, the zinc electrode exhibited a more polished surface and a more compact solid electrolyte interphase (SEI) layer. XRD further supported these findings by affirming the lack of dendritic structures in the zinc deposition with the HE-50 electrolyte. The protective function of the SEI layer was evident as it successfully impeded the formation of zinc dendrites and thwarted unwanted side reactions throughout the processes of zinc stripping and plating in aqueous electrolytes. This, in turn, contributes to an extended operational life for Zn–S batteries [19]. The Zn/50-S@NPC battery, utilizing a hybrid aqueous electrolyte, undergoes an electrochemical conversion reaction that progresses through a succession of consecutive steps. As the discharge progresses, elemental sulfur undergoes a successive reduction, leading to the formation of ZnS_2 and ZnS. These individual transformations are recognizable by unique peaks observed in the Raman spectrum. Throughout the charging process, ZnS undergoes changes, transforming into ZnS_2 and ZnS_x, and subsequently reoxidizes back to elemental sulfur. XPS was employed to validate the reversible chemical transformation occurring between S and ZnS. Although the Raman spectra signaled the appearance and disappearance of diverse sulfur compounds, indicating the presence of polysulfide intermediates throughout the processes of discharging and charging, it is noteworthy that soluble polysulfides were conspicuously absent post-cycling.

The ability of polysulfides to dissolve in both organic and aqueous solvents has positioned them as a central point of interest as potential mediums for energy storage [20, 21]. Noteworthy is the fact that short-chain sulfides often display significant solubility in aqueous environments, a trait that diverges from their performance in organic electrolytes, where short-chain sulfides tend to exhibit insolubility. Therefore, when dealing with significant concentrations of polysulfides or short-chain sulfides within cells, the necessity for surplus electrolytes diminishes when employing solid sulfur cathodes. This feature facilitates the development of systems with elevated energy density. The framework of rechargeable ZAPS batteries has been defined, incorporating essential components, encompassing a zinc metal anode, an alkaline anolyte, an aqueous polysulfide catholyte, a catalytic electrode, and a mediator-ion solid-state electrolyte (SSE) [20]. The essential function of SSE lies in facilitating the transportation of ions from the anolyte to the catholyte.

Illustrated in Fig. 3.3b, the discharge mechanism involves the oxidation of metallic zinc on the anode according to Eq. 2.22. In cases where the polysulfide concentration in the catholyte is substantial, as outlined in Eq. 2.23, the dissolved sulfurs, mainly in the S_4^{2-} form, undergo reduction to S^{2-} at the cathode's catalytic electrode surface. If the concentration of polysulfide is reduced, the process of S_4^{2-} reduction occurs via Eq. 2.24 [22]. It is crucial to highlight that, in the charging process, the reverse reactions take place. The equilibrium reaction for a Zn–S battery operating in an aqueous solution with a moderate polysulfide concentration can be derived through the amalgamation of Eqs. 2.22 and 2.24, as expressed in Eq. 3.1. In Fig. 3.3b, the metal sulfide might incorporate alkali metals, with metallic elements, encompassing Na or Li being potential constituents [23].

$$Zn + 4OH^- \rightarrow [Zn\,(OH)_4]^{2-} + 2e^- \qquad (3.1)$$

$$S_4^{2-} + 6e^- \rightarrow 4S^{2-} \tag{3.2}$$

$$S_4^{2-} + 6e^- + 4H_2O \rightarrow 4HS^- + 4OH^- \tag{3.3}$$

$$3Zn + 8OH^- + S_4^{2-}4H_2O \rightarrow 3[Zn(OH)_4]^{2-} + 4HS^- \tag{3.4}$$

Despite the broad adoption of M–S batteries utilizing organic or non-polar electrolytes, encompassing Li–S and Na–S batteries, due to their remarkable performance, their wider application faces restrictions outlined in Fig. 3.2, attributed to specific drawbacks. The potential hazards and propensity for explosions are common attributes of organic electrolytes. Considering growing concerns regarding battery safety, researchers are increasingly intending to substitute organic electrolytes with aqueous alternatives. Consequently, there is a rising inclination among researchers to opt for secure M–S batteries that incorporate aqueous electrolytes. Researchers are actively working to tackle inherent challenges associated with these batteries, encompassing restrictions in ESPW, irreversibility stemming from reactions involving H_2S and SO_4^{2-}, and sluggish kinetics. This has emerged as a primary area of focus for improvement [12]. The battery's capability for potential output is constrained by the restricted ESPW, which, in turn, excludes specific high-capacity redox pairs within the potential range from OER to HER. Nevertheless, various strategies can be employed to expand the ESPW, encompassing implementing solid electrolyte interphase, regulating the pH of the electrolyte, introducing functional additives, optimizing the electrode structure design, and exploring alternative redox pairs [24]. The occurrence of irreversible side reactions on the electrodes' surfaces poses an extra hurdle for aqueous Zn–S batteries. Specifically, irreversible transformations involving H_2S give rise to the production and release of H_2S gas from the electrolyte. The outflow of this gas leads to a depletion of sulfur, a crucial active material. Additionally, the presence of H_2S has the potential to induce significant corrosion in certain electrodes, particularly stainless-steel electrodes, ultimately resulting in the battery's malfunction [12]. Another notable drawback associated with employing aqueous electrolytes in Zn–S batteries is the irreversible production of SO_4^{2-} throughout the charging process, which stands out as a commonly cited by-product. The produced SO_4^{2-} significantly diminishes the sulfur content within the electrolyte and adversely impacts the reversibility of the electrode in the process [7, 14].

Li et al. [14] conducted an investigation to evaluate the influence of different aqueous electrolytes on the electrochemical efficiency of Zn–S batteries. In this investigation, the anode and cathode electrodes were constituted by zinc metal and 50% sulfur embedded in carbon nanotubes (S@CNTs-50), respectively, alongside different aqueous electrolytes, including $1MZn(CF_3SO_3)_2$, $1MZnSO_4$, and $1MZn(CH_3COO)_2$. Based on the obtained outcomes, depicted in Fig. 3.4, when subjected to a current density of 100 mA/g in the $1MZn(CF_3SO_3)_2$ electrolyte, the S@CNTs-50 cathode exhibits a reversible capacity of 230 mAh/g_S, an initial CE of 63%, and potential hysteresis of 1.1 V. The utilization of $1MZnSO_4$ as the electrolyte in the Zn–S batteries results in a marginal increase in reversible capacity, reaching around 280 mAh/g_S at identical current density. Nevertheless, the contrast between its initial and subsequent cycles highlights an inadequate CE of 26%. Moreover, the battery's voltage hysteresis undergoes a slight decrease, reaching 0.95 V. In the instance of the battery utilizing an electrolyte consisting of 1 M Zn(CH$_3$COO)$_2$, it attains a reversible capacity of approximately 685 mAh/g_S. Alongside its reversible capacity, a substantial increase is evident in the CE, reaching an impressive 98%. Despite the 1 M Zn(CH$_3$COO)$_2$ electrolyte outperforming the other two electrolytes in terms of reversible capacity and CE, a considerable potential hysteresis of 0.9 V is detected. This underscores the need for introducing a suitable additive to enhance the performance of the 1 M Zn(CH$_3$COO)$_2$ electrolyte. Luo et al. [13] explored Zn–S aqueous batteries. The experimental setup featured a zinc foil as the anode, a composite cathode consisting of Ketjen Black-Sulfur (KB-S), and a 1 M ZnCl$_2$ aqueous electrolyte. The attained specific capacity reached 1668 mAh/g_S, while the energy density stood at 1083.3 Wh/kg_S under a current density of 50 mA/g. Yet, owing to the elevated observed potential hysteresis, standing at 0.8 V and associated with the irreversible formation of ZnSO$_4$ [25–27], its CE remains at a modest 30%. Consequently, the recorded discharge capacity undergoes a significant deterioration from 1668 to 560 mAh/g_S after completing four charge and discharge cycles, as depicted in Fig. 3.5a. In a separate investigation performed by Zhang et al. [28], it was revealed that elevating the concentration of ZnCl$_2$ to 30 M resulted in enhanced cycling stability for the manufactured battery. However, this positive effect was accompanied by a substantial reduction in the specific capacity delivered by the battery. The graphical representation in Fig. 3.5b depicts the variation in specific capacities over a span of 50 cycles. As observed in Fig. 3.5, the specific capacity exhibited in the concentration of 30 M ZnCl$_2$ is considerably inferior to that of the battery utilizing 1 M ZnCl$_2$. Additionally, a significant decline is apparent over the span of 50 cycles. Exploratory studies have been undertaken to assess how changes in current density affect the electrochemical performance of the Zn–S battery. As depicted in Fig. 3.5c, the rise in current density, shifting from 50 to 1000 mA/g, leads to an impressive decline in the discharge potential of the KB-S cathode. This phenomenon is attributed to an increased polarization in the interaction between sulfur and zinc ions. Additionally, Fig. 3.5d displays the self-discharge performance, a pivotal metric for Zn–S batteries. As evident, the battery exhibits a modest self-discharge performance three weeks' post-operation, with a measured capacity retention of 80.1%.

Zn(OTF)$_2$ is an alternative electrolyte explored in Zn–S aqueous batteries, as investigated by Xu et al. [15]. The study determined that the optimal concentration for this electrolyte is 3 M, deemed suitable for integration into a Zn–S aqueous battery configuration comprising a cathode with ordered mesoporous carbon embedded in sulfur (CMK-3@S) and a zinc anode. Despite the decline in specific capacities, observed as concentrations of Zn(OTF)$_2$ increase from 1195 mAh/g at 1 M to 926 and 788 mAh/g at 2 and 3 M, in turn, there is a noticeable enhancement in the suppression of side reactions during the charging process. This enhancement contributes to an overall increase in capacity retention. Additionally, within a battery

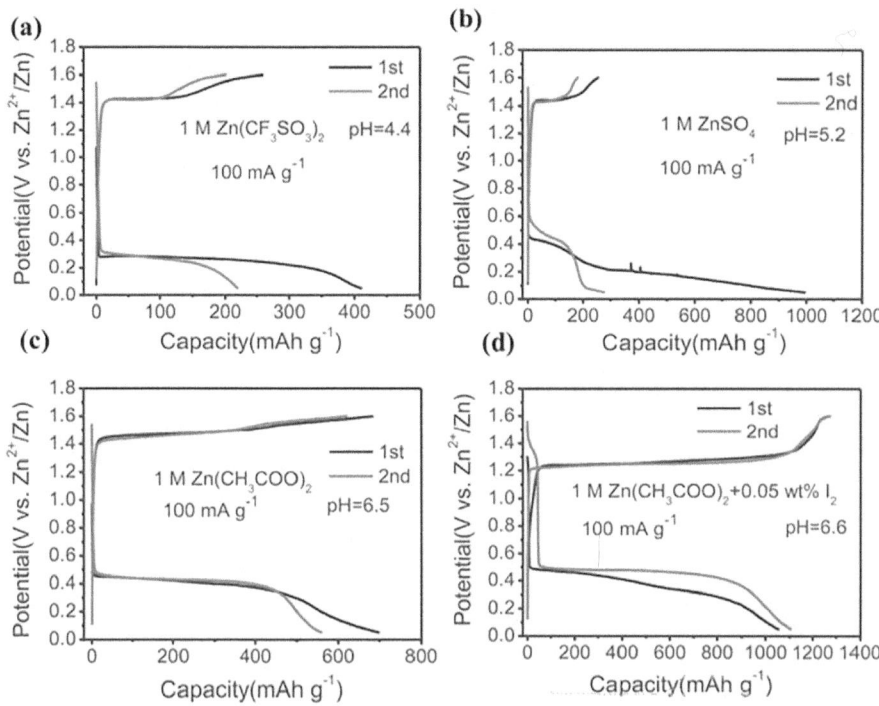

Fig. 3.4 The charge and discharge profiles for the S@CNTs-50 cathode are showcased within aqueous electrolytes, delineated as follows: **a** utilization a 1 **M Zn(CF₃SO₃)₂** solution, **b** employing 1 **M ZnSO₄**, **c** with a 1 **M Zn(CH₃COO)₂** composition, and **d** incorporating 1 **M Zn(CH₃COO)₂** along with an **I₂** additive. Reproduced with permission from [14]

employing a 3 M Zn(OTF)₂ electrolyte, there is an absence of undesirable reactions between water and the electrodes, even at elevated potential. As a result, a significant enhancement in the cycling performance of the battery is anticipated. The zinc electrode's linear polarization curves revealed that augmenting the Zn(OTF)₂ electrolyte concentration resulted in an upsurge in the corrosion potential, $E_{corr.}$, and an associated reduction in the corrosion current density, $I_{corr.}$, of the electrode. These modifications suggest that the zinc electrode experiences diminished dendritic formation in a 3 M Zn(OTF)₂ electrolyte, resulting in reduced corrosion compared to its counterparts with lower concentrations. In their study, Zhao et al. [29] explored the realm of ZPBs. The experimentation incorporated the use of cathodes, including poly Li₂S₆-r-DIB copolymer (PLSD), ionic liquid poly Li₂S₆-r-DIB copolymer (IL-PLSD), and liquid film poly Li₂S₆-r-DIB copolymer (LF-PLSD). These cathodes were combined with a zinc anode and a 1 M Zn(TFSI)₂ electrolyte to construct the ZPBs. The cyclic voltammetry (CV) plots in Fig. 3.6a–c showcase the performance of batteries equipped with PLSD, IL-PLSD, and LF-PLSD cathodes. These plots were achieved at various potential scanning rates: 0.3, 1.0, 3.0, and 5.0 mV/s. Observing Fig. 3.6a reveals the absence of redox peaks, suggesting that the PLSD cathode

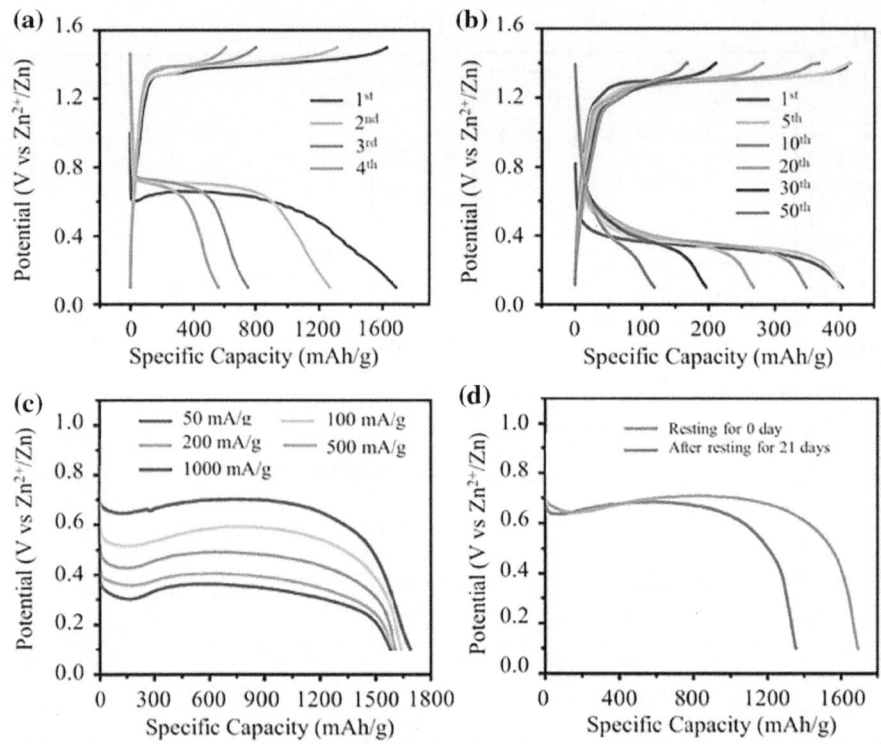

Fig. 3.5 **a** Alternation in the GCD profiles observed after completing four cycles. **b** The GCD plots for the Zn–S battery, which incorporates a 30 M ZnCl$_2$ electrolyte. **c** The KB-S cathode's discharge potential plots under varying current densities in a 1 M ZnCl$_2$ electrolyte. **d** The KB-S cathode's self-discharge performance under a current density of 50 mA/g in a 1 M ZnCl$_2$ electrolyte. Reproduced with permission from [13].

does not partake in reduction and oxidation reactions. The validity of this assertion is also supported by Fig. 3.6d, illustrating the GCD plots of the Zn-PLSD battery throughout three charge and discharge cycles. Conversely, the introduction of IL to PLSD, yielding the IL-PLSD cathode, leads to the emergence of redox peaks in the CV plots, as evident in Fig. 3.6b. As illustrated in Fig. 3.6b, discernible oxidation and reduction peaks appear within the potential intervals of 1.45–175 and 1.14–1.36 V, in turn, across potential scan rates from 0.3 to 5.0 mV/s. Furthermore, elevating the potential scan rate results in a heightened distinction between the observed redox peaks.

The performance over multiple cycles of the Zn/IL-PLSD battery under a current density of 1 A/g is depicted in Fig. 3.7a. In this illustration, it is evident that the discharge capacity undergoes a significant reduction to 126 mAh/g during the initial 100 cycles, yielding a capacity retention of 28%. Continuing cycles up to 464 also lead to a decline in discharge capacity, reaching 30 mAh/g. Additionally, the observed CE values surpassing 100% throughout the 464 cycles are attributed to irreversible

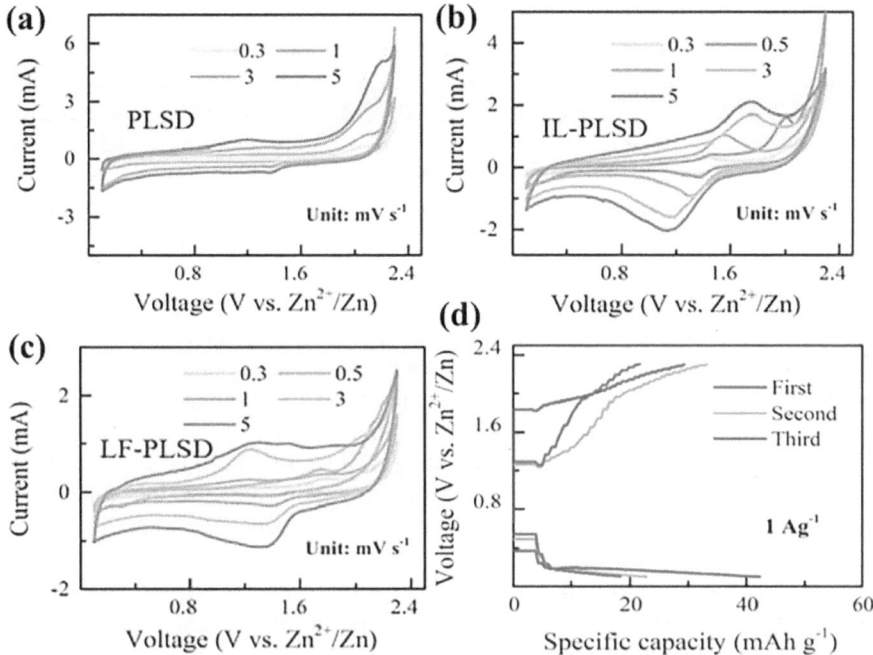

Fig. 3.6 **a–c** The CV plots recorded at scan rates spanning 0.3–5 mV/s, **a** Zn/PLSD, **b** Zn/IL-PLSD, and **c** Zn/LF-PLSD batteries incorporating 1 M Zn(TFSI)$_2$ electrolyte. **d** The GCD plots for Zn/PLSD with 1 M Zn(TFSI)$_2$ electrolyte under a current density of 1 A/g. Reproduced with permission from [29]

side reactions taking place on the electrodes' surface and the dissolution of IL on the PLSD cathode surface. Thus, despite its ability to be recharged, the Zn/IL-PLSD battery exhibits unsatisfactory cyclic performance. As shown in Fig. 3.6c, substituting IL with LF on the PLSD cathode surface leads to a noteworthy improvement in the electrochemical efficiency of the Zn/LF-PLSD battery. Examining the cathodic peaks in Zn/IL-PLSD and Zn/LF-PLSD batteries, there is an observable shift in the potential range of these peaks, moving from 1.14–1.36 V in Zn/IL-PLSD to 1.32–1.39 V in the Zn/LF-PLSD battery. These alternations are ascribed to a reduction in potential polarization and an enhancement in the reversibility of Zn/LF-PLSD battery. The cycling behavior of the Zn/LF-PLSD battery, as depicted in Fig. 3.7b, indicates a more gradual decline when contrasted with the Zn/LF-PLSD battery. As observed, the recorded capacity diminishes to 285 mAh/g after 100 cycles, in contrast to the 126 mAh/g capacity exhibited by the Zn/LF-PLSD battery. Additionally, undergoing a further 600 charge and discharge cycles only leads to a 50 mAh/g reduction in capacity, highlighting a substantial enhancement in the electrochemical performance of the Zn/LF-PLSD battery. The Zn/LF-PLSD battery's ability to handle varying rates was scrutinized, and the outcomes are presented in Fig. 3.7c. An apparent decline in capacity is observed throughout the initial 40 cycles, after which the capacity stabilized and remains relatively constant in the subsequent cycles. The subsequent increment in the current density, from 1 to 5 A/g, signifies a restoration in the discharge capacity of the Zn/LF-PLSD battery. At current densities of 1, 3, and 5 A/g, discharge capacities of 170, 152, and 150 mAh/g are evident in Fig. 3.7c. These discharge capacities translate to capacity retentions of 65%, 113%, and 119%, respectively. Additionally, the CE of the Zn/LF-PLSD battery remains consistent at 100% throughout 700 cycles, primarily due to the robust prevention of irreversible side reactions occurring on the electrode's surface. The examination of Zn/LF-PLSD batteries through GCD test, specifically at current densities ranging from 0.3 to 5 A/g, is portraying in Fig. 3.7d. It is clear from the results that considerable reversible capacities of 1148 and 489 mAh/g were achieved at current densities of 0.3 and 0.5 A/g, respectively. In the subsequent stage of enhancing the electrochemical capabilities of the Zn/LF-PLSD battery, two additional electrolytes were introduced. Specifically, these involved the utilization of 1 M trifluoromethanesulfonate (1 M $Zn(CF_3SO_3)_2$) alongside 21 M LiTFSI, and 1 M $Zn(TFSI)_2$ in conjunction with 1 M LiTFSI. Notably, the primary factors guiding the selection of these electrolytes are the heightened ionic conductivity and electrochemical stability inherent in $CF_3SO_3^-$ and $TFSI^-$ [16].

Analyzing the zinc stripping and plating performances of Zn/LF-PLSD batteries under a current density of 1 mA/cm^2, as depicted in Fig. 3.8a, a comparison was drawn among batteries featuring 1 M $Zn(TFSI)_2$, 1 M $Zn(CF_3SO_3)_2$+1M LiTFSI,

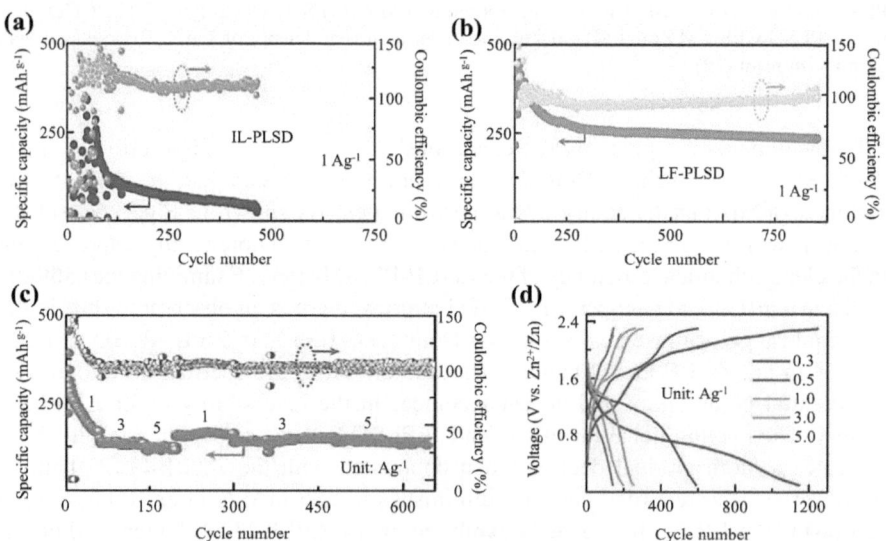

Fig. 3.7 **a, b** The durability graphs for **a** Zn/IL-PLSD and **b** Zn/LF-PLSD batteries, utilizing 1 M $Zn(TFSI)_2$ electrolyte, recorded at a current density of 1 A/g. **c** Assessing the stability under varying rates and **d** examining the GCD graphs of Zn/LF-PLSD with 1 M $Zn(TFSI)_2$ electrolyte. Additionally, scrutinizing the cyclic stability curve for Zn/LF-PLSD batteries incorporating 1 M $Zn(TFSI)_2$ electrolyte. Reproduced with permission from [29]

and 1 M Zn(TFSI)$_2$+1 M LiTFSI electrolytes. This finding indicates that, due to its heightened reversibility and enhanced stability, the Zn/LF-PLSD battery incorporating the 1 M Zn(CF$_3$SO$_3$)$_2$+21 M LiTFSI electrolyte showcase the most favorable electrochemical efficiency, establishing it as the optimum configuration for the Zn/LF-PLSD battery. The CV assessment, conducted at varying scan rates (0.3, 0.5, 1, and 3 mV/s), is visually represented in Fig. 3.8b, a comparative analysis of the redox peak potentials derived from the Zn/LF-PLSD battery with 1 M Zn(TFSI)$_2$ and the corresponding reduction and oxidation peaks observed in the Zn/LF-PLSD battery with 1 M Zn(TFSI)$_2$+1 M LiTFSI electrolyte, situated at 1.47–1.5 V and 1.6–1.64 V, respectively, serves as evidence supporting the improved reversibility in the Zn/LF-PLSD battery incorporating 1 M Zn(TFSI)$_2$+1 M LiTFSI electrolyte. Examining the electrochemical impedance spectroscopy (EIS) outcomes displayed in Fig. 3.8c, it becomes apparent that the Zn/LF-PLSD battery featuring 1 M Zn(TFSI)$_2$+1 M LiTFSI electrolyte exhibits an increased charge-transfer resistance (R_{ct}) in comparison with the Zn/LF-PLSD battery incorporating 1 M Zn(TFSI)$_2$.

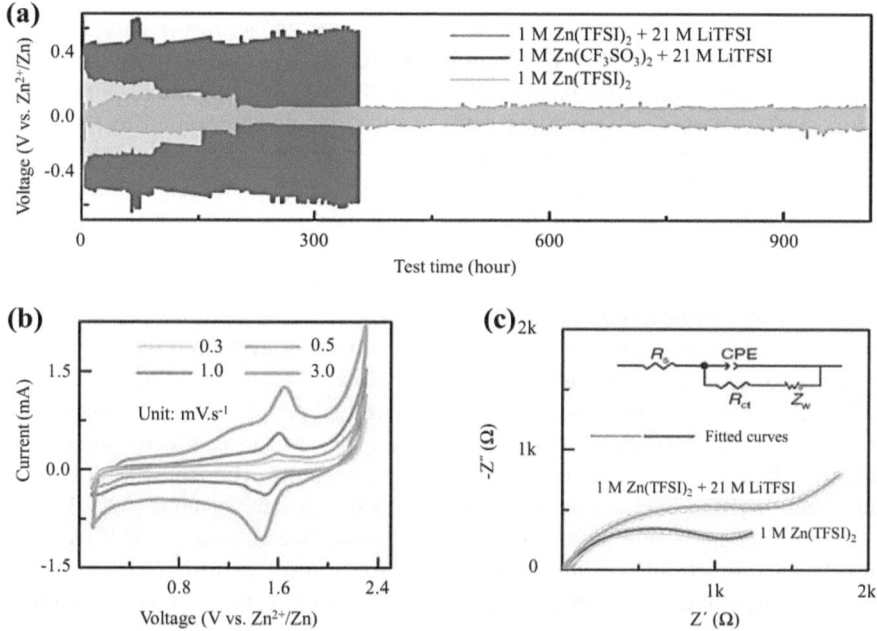

Fig. 3.8 **a** The zinc stripping and plating graphs for Zn/LF-PLSD batteries, employing 1 M Zn(TFSI)$_2$, 1 M Zn(CF$_3$SO$_3$)$_2$ +21 M LiTFSI, and 1 M Zn(TFSI)$_2$ + 1 M LiTFSI electrolytes. **b** The CV graphs for Zn/LF-PLSD battery, utilizing 1 M Zn(TFSI)$_2$ + 1 M LiTFSI electrolyte, recorded at scan rates ranging from 0.3 to 3.0 mV/s. **c** The EIS plots derived from Zn/LF-PLSD batteries incorporating 1 M Zn(TFSI)$_2$ + 1 M LiTFSI and 1 M Zn(CF$_3$SO$_3$)$_2$ +21 M LiTFSI electrolytes, along with their corresponding equivalent circuit. Reproduced with permission from [29]

In a separate investigation, Cai et al. [30] endeavored to confront multiple hurdles by combining an alkali zinc anode with an acidic sulfur electrode. In the pursuit of this objective, a sulfur host is engineered through the dispersion of atomic Zn–N$_4$ on nitrogen-doped hollow porous carbon (Zn–NHPC). This design boosts efficiency by capitalizing on the superior affinity of Zn–N$_4$ for CuS in comparison with N-doped graphene. Consequently, this design minimizes the excessively high barrier for vulcanization reactions that is typically associated with N-doped graphene. The hybrid Zn–S battery showcased a substantial 380% increase in energy density, outperforming the maximum values documented by other researchers [13]. This highlights the crucial aspect of concurrently refining the reaction conditions for both zinc anode and sulfur cathode. By doing so, it opens a potential avenue for improving the kinetics and rate performance of the battery reactions.

3.2.2 Influence of Redox Mediators in Aqueous Electrolytes

Serving as a versatile electrolyte additive, redox mediators play a pivotal role in engaging in reversible redox reactions. Their primary impact lies in enhancing the ZnS–S reaction's reversibility, resulting in the provision of extra capacities to Zn–S batteries through diverse mechanisms. In the realm of both Zn–S and Li–S batteries, redox mediators assume individual roles, each meticulously crafted to overcome challenges inherent to their corresponding battery systems. Redox mediators play a crucial role in addressing challenges inherent to sulfur's insulating characteristics in Li–S batteries. They assist in transforming lithium polysulfides and minimizing the shuttle effect, a complication less commonly observed in Zn–S batteries [15, 31]. Contrary, redox mediators in Zn–S batteries concentrate on accelerating the kinetics of the sulfur redox reaction and fortifying cycling stability, particularly in response to the substantial expansion in sulfur volume throughout the discharging process [18, 23]. Notwithstanding these operational variations, redox mediators in both battery types are formulated to be regenerative, partaking in reversible redox processes throughout both discharge and charge cycles. Ensuring effectiveness across numerous battery cycles, this regeneration plays a vital role in boosting the efficiency, rate capability, and durability of both Li–S and Zn–S battery systems [32, 33].

Examining Zn–S batteries, Chang et al. [7] explored the influence of TU, a recently introduced bifunctional additive, on the electrochemical performance of a Zn–S battery configuration. This setup comprised a zinc anode, cathodes constructed from sulfur embedded in Ketjen Black (S@KB) and KB, and utilized ZnSO$_4$ as the electrolyte. The disclosure from their study indicates that introducing 2 g/L TU results in a further boost to the battery's capacity, an enhancement in the electrochemical reactions, and consequently, an overall improvement in kinetics. While undergoing the charging process, Thiourea ion (TU$^-$) derived from TU reacts with the electrons originating from nitrogen, sulfur atoms, and zinc ions produced according to Eq. 3.5 at the surface of the zinc anode. This process causes a decrease in the bonding energy of the ZnS bond.

$$2ZnS + 4H_2O - 10e^- \rightarrow 2Zn^{2+} + SO_4^{2-} + S + 8H^+ \tag{3.5}$$

On the flip side, throughout the charging phase, the metallic ions generated can engage with sulfur atoms, leading to an additional decrease in the bonding energy of the ZnS bond. The collaborative effect of these two reactions expedites the production of zinc ions, facilitating their diffusion within the electrolyte. The manifestation of this process can be observed in Fig. 3.9a, b. The diffusion coefficient stands as a factor influencing ion mobility. Electrolytes boasting higher diffusion coefficients contribute to batteries experiencing faster reactions. The quantities of diffusion coefficients during the charging and discharging phases in both electrolytes (with and without the TU additive) are presented in Fig. 3.9a, b. Notably, in Fig. 3.9a, the absence of TU in the electrolyte results in a considerably lower diffusion coefficient during the charging phase compared to the discharging phase. This observed variation leads to a significant impediment in the reaction kinetics on the cathode during the charging phase. Conversely, the incorporation of TU into the electrolyte brings about a remarkable enhancement in the diffusion coefficient during charging, thereby expediting the cathodic reaction, as depicted in Fig. 3.9b. The data acquired from EIS in Fig. 3.9c serves as additional evidence supporting the previously mentioned assertions. Moreover, the inclusion of TU in the electrolyte enhances the reactivity between carbonium ions and sulfur atoms of ZnS throughout the charging process. Based on the descriptions provided, the incorporation of TU stands out as a potential avenue for enhancing the electrochemical performance of Zn–S aqueous batteries. In Fig. 3.9d, the CV curves for $KB/ZnSO_4$, $KB/ZnSO_4 + TU$, $S@KB/ZnSO_4$, and $S@KB/ZnSO_4 + TU$ are depicted. Notably, no discernible peaks are evident in the case of $KB/ZnSO_4$. However, the inclusion of TU in the electrolyte results in the emergence of two peaks at 0.38V and 1.18 V in the reduction branch, along with a peak at 1.24 V in the oxidation branch. The occurrence of two reduction peaks at 0.38 V and 1.18 V can be ascribed to the formation of ZnS and the reduction of TU, respectively. Similarly, the oxidation peak is associated with the transformation of ZnS into S, as outlined in Eq. 2.6.

Contrasting the CV curves derived from $KB/ZnSO_4$ and $KB/ZnSO_4 + TU$ suggests that throughout the charging and discharging cycles, the additive TU actively engages in the redox reaction, thereby amplifying the specific capacity of the Zn–S battery. This analogous phenomenon is also observed in the cathode of S@KB, as depicted in Fig. 3.9d. It is noteworthy to highlight that for the S@KB cathode, the introduction of TU into the electrolyte results in a subtle shift in the reduction peaks. Nevertheless, the potential associated with the oxidation peak, linked to the conversion of ZnS–S, shifts from 1.64V to 1.24 V. This indicates a considerable reduction in the energy barrier and, consequently, an enhancement in the kinetics of the ZnS–S conversion process.

Examining the Galvanostatic Intermittent Titration Techniques (GITT) curves presented in Fig. 3.10a, b, it is evident that the Zn–S batteries, when operated without TU, exhibit charging and discharging capacities of 2193.8 mAh/g and 1531.2 mAh/g, respectively. However, the utilization of TU increases these capacities to around 2032.1 mAh/g. As previously noted, a prevalent drawback encountered in aqueous

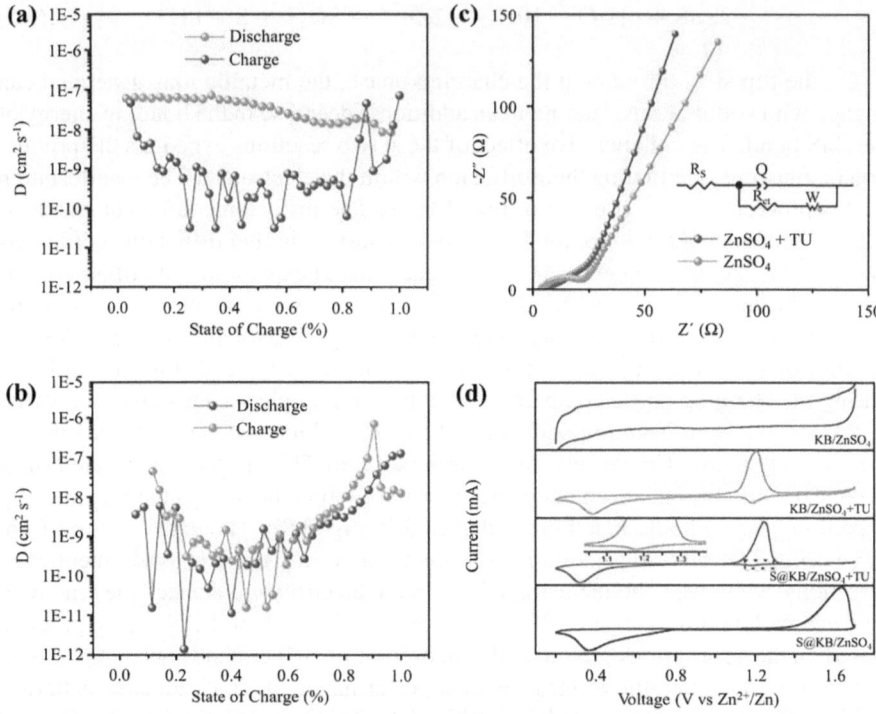

Fig. 3.9 a, b Diffusion coefficient of the S^{2+} ion in **a** $ZnSO_4$ electrolyte, **b** $ZnSO_4$ electrolyte with the addition of TU. **c** Nyquist plots illustrate the behavior of S@KB cathode in $ZnSO_4$ and $ZnSO_4 + TU$ electrolytic environments. **d** The CV profiles correspond to $KB/ZnSO_4$, $KB/ZnSO_4 + TU$, $S@KB/ZnSO_4$, and $S@KB/ZnSO_4 + TU$. Reproduced with permission from [7]

Zn–S batteries involves the production of SO_4^{2-} side-products during the charging process. The occurrence leads to a depletion of active sulfur within the electrolyte, consequently causing a substantial decrease in the reversibility of the electrode [7,14]. The validation of this assertion is evident in the comparison of charging capacities depicted in Fig. 3.10a, b. It substantiates that the introduction of TU to the $ZnSO_4$ electrolyte effectively hinders the formation of SO_4^{2-}. It is important to note that the increased charging capacity observed in the battery using TU-free electrolyte is a consequence of the formation of the SO_4^{2-} by-product. As per Eq. 3.5, the generation of SO_4^{2-} entails the transfer of eight electrons, whereas only two electrons are involved in the conversion of ZnS–S, as outlined in Eqs. 2.4–2.6. On the flip side, the augmented discharge capacity observed in the battery featuring TU in its electrolyte points to an intensified redox reaction, as evidenced by the CV curves. Moreover, the substantial decrease in voltage hysteresis, reducing from 0.81 V to 0.46 V in Fig. 3.10a, b are a direct outcome of introducing TU to the $ZnSO_4$ electrolyte. The diminishing value reflects a reduction in the reaction polarization of the battery when TU is incorporated.

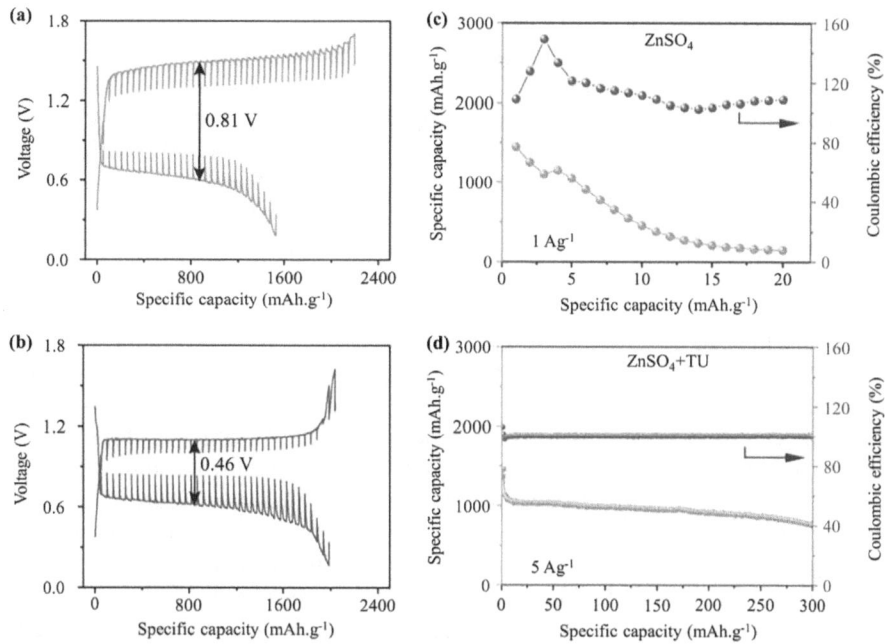

Fig. 3.10 **a, b** GITT profiles for the S@KB cathode in **a** ZnSO$_4$ electrolyte, **b** ZnSO$_4$ electrolyte with the addition of TU. **c, d** Specific capacity and CE of S@KB cathode in **c** ZnSO$_4$ electrolyte and **d** ZnSO$_4$ + TU electrolyte. Reproduced with permission from [7]

Figure 3.10c, d depict alterations in the specific capacity and CE of the S@KB cathode throughout battery cycling in ZnSO$_4$ and ZnSO$_4$ + TU, respectively. When employing a current density of 1 A/g, it becomes evident that the CE of the S@KB cathode exceeds 100% during cycling, while its specific capacity undergoes a pronounced decline after 20 cycles. This underscores that the elevated CE is ascribed to the creation of the by-product SO$_4^{2-}$. On the flip side, the inclusion of TU in the electrolyte proves effective in mitigating capacity deterioration, with only a marginal 0.11% decline per cycle observed over 300 cycles at a current density of 5 A/g. Additionally, the CE remains consistently close to 100%, owing to the suppression of SO$_4^{2-}$ by-product formation.

Liu et al. [3] explored the impact of the TUI redox mediator on the Zn–S aqueous battery's performance, featuring a carbon fiber reinforced ZnS (ZnS@CF) cathode. Through the addition of 1 wt% TUI to the 3 M ZnSO$_4$ aqueous electrolyte, and subsequent oxidation by I$_2$, they observed the formation of 2TU^{2+} and 2I$^-$. Considered an efficient mediator, I$^-$ plays a crucial role in minimizing overpotential during the charging process. Figures 3.11a and 3.10b reveal a substantial reduction in the average overpotential during both the initial and subsequent charges, dropping from approximately 1.6 V to nearly 1.2 V. The decrease observed signifies that the inclusion of TUI lowers the activation energy associated with the conversion of ZnS to S, thereby promoting the reaction. Additionally, reports are indicating that TU hinders the formation of dendritic structures on the surface of the zinc electrode. Such structures can be detrimental to the cycling performance of Zn–S batteries. Moreover, TU enhances the corrosion resistance of the electrode. Figure 3.11a, b clearly demonstrate that introducing TUI brings about a significant increase in discharge capacity, improving it from 216 mAh/g to almost 480 mAh/g during the second cycle. Furthermore, the discharge capacity experiences a noticeable decline from the first to the second cycle when the electrolyte is lacking TUI, indicating a lack of robust stability in the battery. Conversely, the inclusion of TUI leads to a notable enhancement in stability. As illustrated in Fig. 3.11c, there is marginal alteration in the charge and discharge capacities when the number of charge and discharge cycles is increased to 5 and 10. This can be attributed to the exceptional cycle stability demonstrated by the Zn–S battery, featuring 3 M $ZnSO_4$ + 1wt.% TUI. Alongside the amelioration in discharge capacity, TUI plays a role in diminishing voltage hysteresis, reducing it from 1.03 V to 0.72 V in the first cycle and from 0.89V to 0.5 V in the second cycle. The outcomes observed in charge and discharge experiments align with those obtained from CV tests, reinforcing the validity of the findings. Figure 3.11d illustrates the absence of a discernible peak in the oxidation branch of the battery when utilizing a TUI-free electrolyte. This suggests that the oxidation reaction of ZnS is not viable within this specific potential range. Conversely, the inclusion of TUI introduces a cathodic peak at 1.2 V, indicating the occurrence of oxidation for both ZnS and I^-. The reduction peak identified at 1.21 V is linked to the reduction of I_3^-, whereas the two additional reduction peaks are indicative of the formation of ZnS.

Liu et al. [3] also employed ZnI_2 as a redox mediator and juxtaposed its performance with that of TUI. Figure 3.12a displays a charge–discharge plot that closely resembles the electrochemical behavior observed with the TUI additive. This similarity provides further confirmation of the impact of I^- on the redox reaction and specific capacity. A thorough examination of the impact of TUI and ZnI_2 redox mediators on capacity retention and CE over 100 cycles, as depicted in Fig. 3.12b, reveals a uniform and comparable CE for both batteries throughout the cycling process, maintaining a high level of around 100%. Additionally, while the battery featuring an electrolyte with the TUI additive maintains nearly 100% capacity retention, the battery with the ZnI_2 redox mediator experiences a significant capacity decay of 35%. In Fig. 3.12c, the diffusion coefficient of the ZnS@CF cathode is showcased within the 3 M $ZnSO_4$ electrolyte, both in the absence and presence of 1 wt% TUI, as monitored through GITT. The introduction of TUI is evident in facilitating ion diffusion. Furthermore, the inclusion of TUI enhances the diffusion coefficient of ZnS@CF in the second cycle when compared to its performance in the initial cycle.

In their findings, Liu et al. [3] demonstrated that the Zn–S battery, composed of a zinc anode, ZnS@CF cathode, and $ZnSO_4$ + TUI electrolyte, exhibits superior electrochemical performance. This includes a specific capacity of 465 mAh/g_{ZnS}, equivalent to 1410 mAh/g_S, and an energy density of 274 Wh/kg_{ZnS} (832 Wh/kg_S) at a current density of 0.1 A/g. In a separate study, investigating the influence of a 0.05 wt.% I_2 redox mediator, Li et al. [14] examined the electrochemical behavior of a Zn–S battery. The battery design comprises a zinc metal anode, utilizing an electrolyte of 1 M zinc acetate (1 M $Zn(CH_3COO)_2$) with a pH of 6.5, and features an

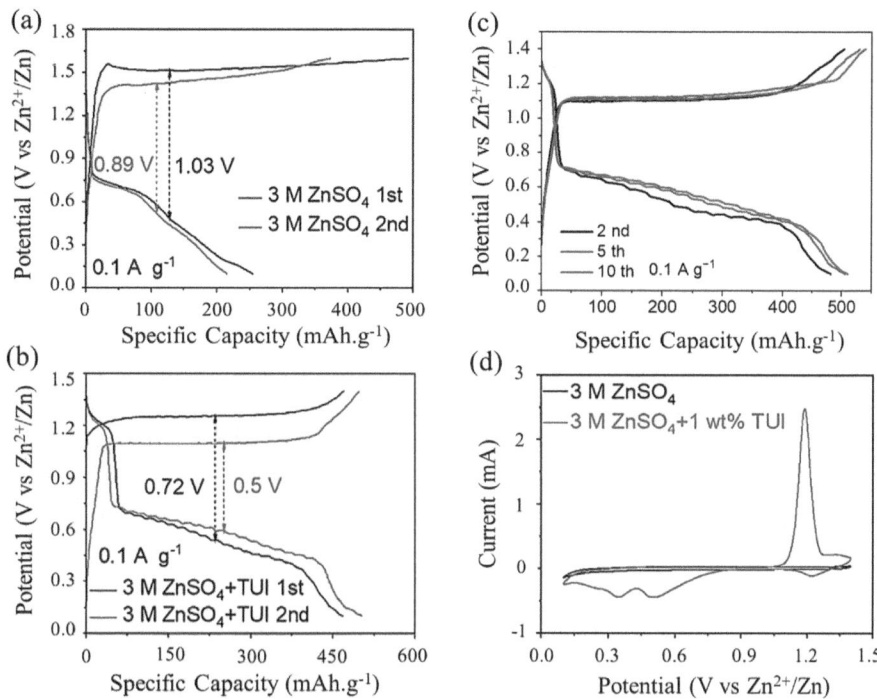

Fig. 3.11 a, b Potential profiles throughout the initial and subsequent charge and discharge processes **a** without TUI, **b** with TUI. **c** The ZnS@CF cathode's charge and discharge characteristics in 3MZnSO₄. **d** CV plots for ZnS@CF cathode in 3MZnSO₄ electrolyte comparing the electrochemical performance in the presence and absence of TUI redox mediator, utilizing a scan rate of 0.1 mV/s. Reproduced with permission from [3]

S@CNTs-50 cathode. Figure 3.4c displays the charge and discharge profiles. Clear in the figure is the remarkable performance of the S@CNTs-50 cathode, exhibiting a reversible capacity of 685 mAh/g, a CE of 98%, and a potential hysteresis of 0.9 V at a current density of 0.1 mA/g. In a bid to improve the electrochemical performance of the battery, the electrolyte was enhanced by the addition of a redox mediator, specifically 0.05 wt.% I_2. In Fig. 3.4d, a notable augmentation is evident in the cathode's reversible capacity, reaching 1105 mAh/g at the identical current density. An energy density of 502 Wh/kg$_S$ is also achieved. Additionally, the observed decline in potential hysteresis to 0.72 V for S@CNTs-50 indicates that the inclusion of I_2 additives promote ion movement, thereby enhancing the overall kinetics of the reactions. Consequently, the improvement in potential hysteresis is a notable outcome of incorporating the I_2 additive. Examining the potential hysteresis in Fig. 3.4c, d reveals a notable drop, moving from 0.9 V to 0.72 V. This decline is associated with a diminish in the polarization of the reaction due to the introduction of the I_2 redox mediator. Furthermore, Fig. 3.12d illustrates the stabilization of plating and stripping reactions on the zinc anode surface achieved by introducing 0.05 wt.% I_2 into the

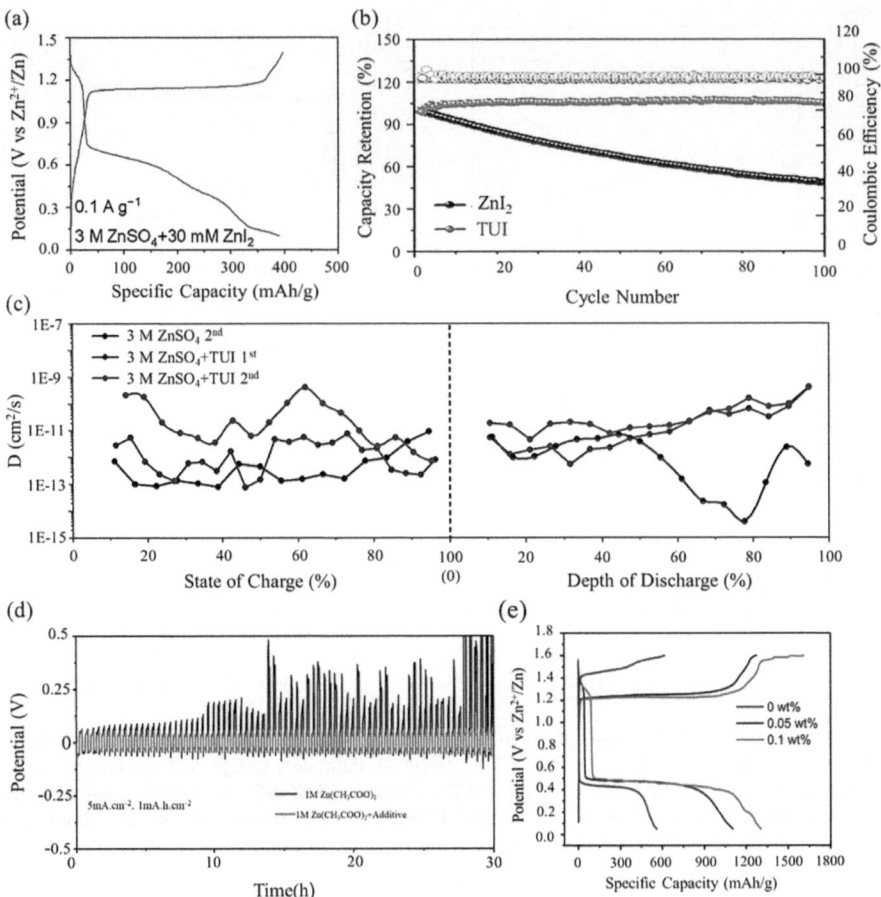

Fig. 3.12 **a** The ZnS@CF cathode's charge and discharge behavior in an electrolyte composed of 3MZnSO$_4$ + 30MZnCl$_2$. **b** The assessment of the capacity ratio and CE for ZnS@CF cathode, utilizing electrolyte containing TUI and ZnI$_2$ redox mediators at a current density of 1 A/g. **c** The diffusion coefficient of zinc ions in the ZnS@CF cathode in both 3MZnSO$_4$ electrolyte and 3MZnSO$_4$ with the addition of 1wt.% TUI [3]. **d** The cycling performance of zinc anode in the electrolyte of 1 M Zn(CH$_3$COO)$_2$ with and without I$_2$ redox mediator. **e** The charging and discharging patterns of S@CNTs-50 in an electrolyte containing 1 M Zn(CH$_3$COO)$_2$ with different concentration of I$_2$ additive [14]. Reproduced with permission from [3, 14]

1 M Zn(CH$_3$COO)$_2$ electrolyte. As depicted in Fig. 3.12e, elevating the concentration of the I$_2$ redox mediator to 0.1 wt.% amplifies the recorded discharge capacity (1302 mAh/g$_S$). However, there are no discernible alternation in the discharge plateau and overpotential. This indicates that an excess of I$_2$ has a marginal effect on overpotential, and the heightened discharge capacity is primarily ascribed to the contribution of I$_2$.

In an effort to overcome the sluggish redox kinetics associated with the ZnS solid–solid conversion, Wu et al. [34] experimented with the incorporation of a dual mediator, specifically trimethylphenylammonium iodide ($Me_3PhN^+I^-$). Me_3PhN^+ serves as a dissolution mediator, facilitating the formation of soluble polysulfide intermediates during the discharging process. This establishes an efficient solid–liquid–solid pathway for sulfur. In the charging process, the oxidation of ZnS to S is facilitated by the I^-/I_3^- redox mediator. Furthermore, the reduced solubility of I_3^- influenced by Me_3PhN^+, effectively prevents undesired shuttling while preserving its catalytic activity.

3.2.3 Gel Electrolytes and Investigating the Influence of Redox Mediators

A prevalent drawback observed in aqueous Zn–S batteries pertains to their performance and adaptability across diverse environmental conditions, notably within a constrained operating temperature range. To overcome this limitation, freeze-resistant hydrogel electrolytes have been implemented. In their work, Amiri et al. [33] engineered an aqueous quasi-solid-state battery configuration. This setup includes a zinc anode, a cathode made of high-strength porous carbon nanofibers enriched with sulfur (CNF-S), and a freeze-resistant electrolyte. The electrolyte formulation includes 1 M $Zn(CH_3COO)_2$, EG serving as the freeze-proofing agent, along with the redox mediator of I_2. To examine the operational characteristics of the developed battery using $Zn(CH_3COO)_2$ electrolyte and investigate the effect of adding 0.1 wt.% I_2 redox mediator, GCD tests were executed at a room temperature of $+25\,°C$ with a current density of 500 mA/g. The results obtained were subsequently compared with two distinctive batteries employing $Zn(CF_3SO_3)_2$ and $ZnSO_4$ electrolytes at the same concentration. As depicted in Fig. 3.13a, while the use of EG electrolyte exhibits a minor advantage over $Zn(CF_3SO_3)_2$ and $ZnSO_4$ electrolytes, the introduction of 0.1 wt.% I_2 to the EG electrolyte significantly boosts the discharge capacity. The discharge capacity rises from 375 mAh/g_{CNF-S} to 667 mAh/g_{CNF-S}, attributed to the impact of the redox mediator. Furthermore, the incorporation of the I_2 redox mediator leads to a decrease in potential hysteresis. As depicted in Fig. 3.13b, the introduction of the I_2 redox mediator brings about a diminish in charge polarization and activation energy, thereby enhancing the electrochemical performance of the batteries under investigation. Furthermore, the presence of the I_2 redox mediator contributes to an improvement in the diffusion coefficients of zinc ions. As depicted in Fig. 3.13c, while undergoing charging process, the introduction of I_2 redox mediator results in an augmentation of diffusion coefficients within the cathode. This transition is noticeable as the values shift from the initial range of 6.5×10^{-11} to 1×10^{-10} cm^2/s to a revised range of 9.5×10^{-11} to 1.46×10^{-10} cm^2/s.

To identify the optimal quantity of EG, discharge tests were performed on zinc acetate batteries that incorporated different concentrations of EG. The experiments

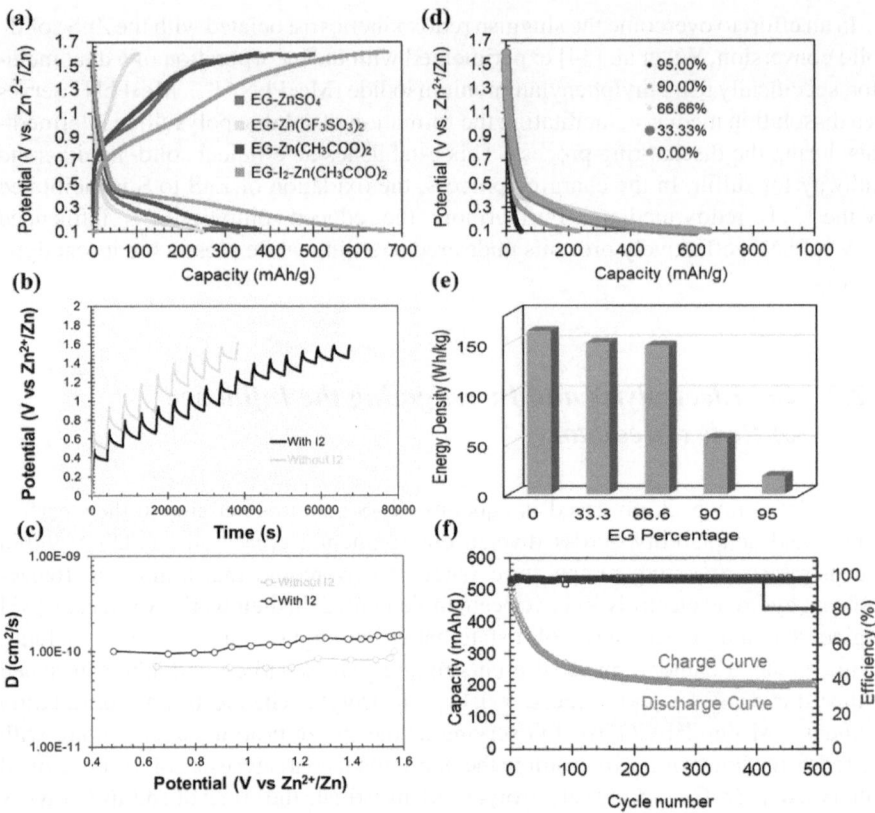

Fig. 3.13 **a** GCD profiles observed in Zn–S batteries with various electrolyte compositions. **b** GITT profiles of Zn–S batteries during the charging process, comparing cases with and without the addition of I₂ redox mediator. **c** Change in the diffusion coefficients of zinc ions within CNF-S cathode as a result of the charging process, **d, e** profiles of **d** discharge and **e** energy density in Zn–S batteries incorporating acetate electrolyte and I₂ redox mediator across different concentrations of EG. **f** Charge and discharge patterns along with CE plots for the Zn–S battery operating at a current density of 1 A/g. Reproduced with permission from [33]

were conducted at 25 °C within the potential range of 0.1–1.6 V. The outcomes of these tests are illustrated in Fig. 3.13d, e. The most effective concentration of EG is identified as 66.6 wt.%, resulting in a marginal decrease of less than 10% in energy density. The battery's cycling performance was then evaluated over 500 cycles at a current density of 1 A/g, incorporating the acetate electrolyte with 66.6 wt.% EG and 0.1 wt.% I₂ redox mediator. The outcome of these evaluations is illustrated in Fig. 3.13f. Throughout the initial 300 cycles, a substantial decline in capacity is noticeable, attributed to irreversible side reactions that involve the generation of hydrogen and oxygen, as well as the formation and release of H₂S from the electrolyte, ultimately causing sulfur depletion. While, in the subsequent 200 cycles, the involvement of the porous CNF-S cathode, comprising sulfur embedded in carbon nanofibers, makes a significant contribution, leading to stabilization in the side reactions and resulting in only a 3% observable capacity decay. Additionally, by the end of 500 cycles, a CE close to 100% is reached.

Amiri et al. [4] engineered a Zn–S battery with inspiration drawn from Kirigami structures. This innovative design integrates cathode material in the form of sulfur-added activated carbon (AC-S) and anode material comprising zinc nanoparticles with an average diameter of 100 nm. The battery utilizes Ag/silicone composite current collectors, as visually depicted in Fig. 3.14a. The electrodes, prepared in their as-is state, underwent a coating process with an anti-freezing gel electrolyte to fabricate a 2D planar Zn–S device. Subsequently, a laser-cutting procedure was applied to cut the device into a Zn–S battery inspired by Kirigami, featuring unique and personalized shapes. Exploring the influence of tensile strain on the electrochemical attributes of the adhesive electrolyte, the specimen experienced distinct strains: 0, 50, 100, 150, and 200%. The discharge profiles and corresponding electrochemical features can be observed in Fig. 3.14b. Conducting adhesive tensile tests aimed at exploring the influence of zinc salts on adhesion strength, pull tests were employed, and the results are presented in Figs. 3.14c and 3.19d. The observed adhesion strength hierarchy in the presence of different zinc salts was determined as $Zn(CF_3SO_3)_2$ > ZnBF> ZnAC − ZnBF (molar ratio of 1 : 1) ≫ ZnAc > $ZnSO_4$. In contrast to the gel electrolytes containing $ZnSO_4$ or ZnAc, which demonstrated weak adhesion strength, the gel electrolyte incorporating ZnAC − ZnBF represented a substantially higher adhesion strength of 33.5 kPa. Despite the superior electrochemical performance of the ZnAC − EG/I_2 hydrogel sample (Fig. 3.14e), it faced challenges in adhesion, experiencing delamination upon exposure to stretching and bending cycles. The hydrogel electrolyte, whether ZnBF or $Zn(CF_3SO_3)_2 − EG/I_2$, demonstrated impressive adhesive strength but presented suboptimal electrochemical features. Seeking to reconcile this compromise, a hybrid strategy involving the synergistic use of ZnAc and ZnBF was adopted. The empirical findings indicatedthat the electrolytes with a molar ratio of 1:1 for ZnAc: ZnBF displayed superior capacity and energy density compared to ZnBF alone (Fig. 3.14e). Additionally, they demonstrated commendable adhesion strength. While the electrochemical properties of various synthesized hydrogel electrolytes were found to be subpar, the use of EG as an anti-freezing agent and I_2 as a redox mediator had no detrimental impact on the performance of the electrolytes.

3.2.4 Organic Electrolytes

While Zn–S batteries derive various advantages from the use of aqueous electrolytes, they are accompanied by drawbacks such as limited ESPW, the emergence of dendritic structures on electrodes, the generation of undesirable by-products resulting from the presence of H_2O in the electrolytes, ultimately causing disturbances in CE and a shortened longevity. To address these issues, one potential

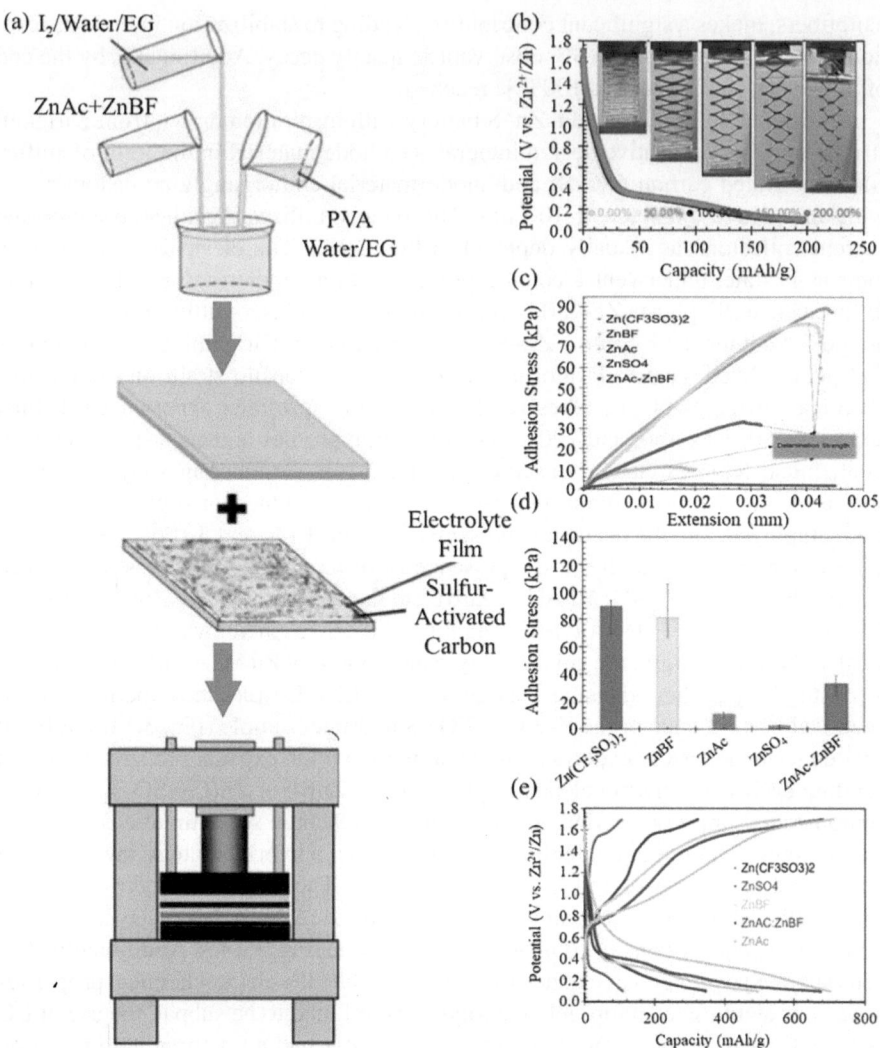

Fig. 3.14   a Schematic representation depicting the fabrication procedure for the 2D planar Zn–S device. **b** Discharge characteristics of the Kirigami-inspired Zn–S battery, illustrating its performance across a range of tensile strains at 25 °C. **c** Stress-extension profiles illustrate unique attributes, **d** measurements for adhesion strength for various hydrogels incorporating zinc salts, and **e** charging and discharging patterns of the Zn–S battery were monitored using various electrolytes at a current density of 0.25 A/g. Reproduced with permission from [4]

approach is the adoption of organic electrolytes, accompanied by careful consideration of safety measures concerning aspects such as flammability and toxicity. To eliminate undesired water-based side reactions, a viable alternative is the incorporation of ionic liquids (ILs). Additionally, employing IL electrolytes that include zinc salts helps suppress hydrogen reactions and the formation of zinc dendrites, presenting a promising solution. Achieving a near-perfect CE of 99% is possible by effectively eliminating dendritic structures [35].

Cui et al. [8] have crafted a DES by combining budget-friendly urea and ChCl as electrolytes. This electrolyte exhibits favorable solubility across a broad spectrum of zinc salts, encompassing ZnS, zinc chloride, zinc nitrate, and zinc acetate. The incorporation of Cl^- serves to reduce electrolyte polarization and enhance ion conductivity [10]. Consequently, $ZnCl_2$ and LiCl were chosen as the zinc salts for the development of the electrolytes. To fine-tune the electrochemical performance and ion conductivity, the incorporation of lithium cations (Li^+) and acetonitrile (AN) as additives was implemented. To summarize the composition of the developed electrolytes, D–Zn, D–Zn–Li, D–Zn–Li–5% AN, D–Zn–Li–10% AN, and 1 M $ZnSO_4$ in water (W–Zn) consist of 1 M $ZnCl_2$, 0.5 M $ZnCl_2$ + 0.5 M LiCl, 0.5 M $ZnCl_2$ + 0.5 M $ZnCl_2$ + 0.5 M LiCl +5 vol.% AN, 0.5 M $ZnCl_2$ + 0.5 M LiCl + 10 vol.% AN, and W-Zn, respectively. As depicted in Fig. 3.15a, the CV profiles of each sample demonstrate analogous patterns, reflecting consistent zinc stripping and plating reactions. However, the distinctive feature is the highest current peak in D–Zn–Li–10% AN, signifying efficient zinc-ion mobility within the electrolyte and enhanced reversibility. These outcomes can primarily be attributed to a substantial reduction in the electrolyte's viscosity resulting from the addition of AN. This results in a notable enhancement in ion conductivity, increasing from 1.23 to 4.66 mS/cm.

A significant drawback of aqueous-based electrolytes lies in their elevated water evaporation rate. Cui et al. [8] noted that in W–Zn batteries, the complete evaporation of water takes forty days. In contrast, the weight retention data depicted in Fig. 3.15b highlights that the utilization of DES effectively addresses and mitigates concerns related to evaporation. The noted upticks in the weight of electrolytes based on DES may be ascribed to the absorption of ambient moisture over a forty-day span. Furthermore, as indicated in Fig. 3.15c, the inclusion of AN in the battery's electrolyte impedes the development of dendritic structures on electrode surfaces, resulting in an improvement in corrosion behavior. In the polarization tests carried out on zinc electrodes for both D–Zn–Li–5% AN and W–Zn, there is a discernible reduction of 0.781 mA/cm^2 in the corrosion current density observed in D–Zn–Li–5% AN when compared to the W–Zn electrolyte [36]. The reduction in electrode corrosion rates within non-aqueous electrolytes that include AN has facilitated the development of batteries characterized by prolonged longevity. As evident in Fig. 3.15d, the use of a 1 M $ZnSO_4$ aqueous electrolyte reveals a restricted ESPW of 3.05 V. The constrained potential creates a favorable environment for unwanted reactions to take place. Thus, to mitigate side reactions and enhance the electrochemical performance of batteries, it becomes essential to widen their ESPW.

Figure 3.15d additionally demonstrates that substituting the electrolyte containing $ZnCl_2$ with a 1 M $ZnSO_4$ aqueous electrolyte results in an expansion of the ESPW

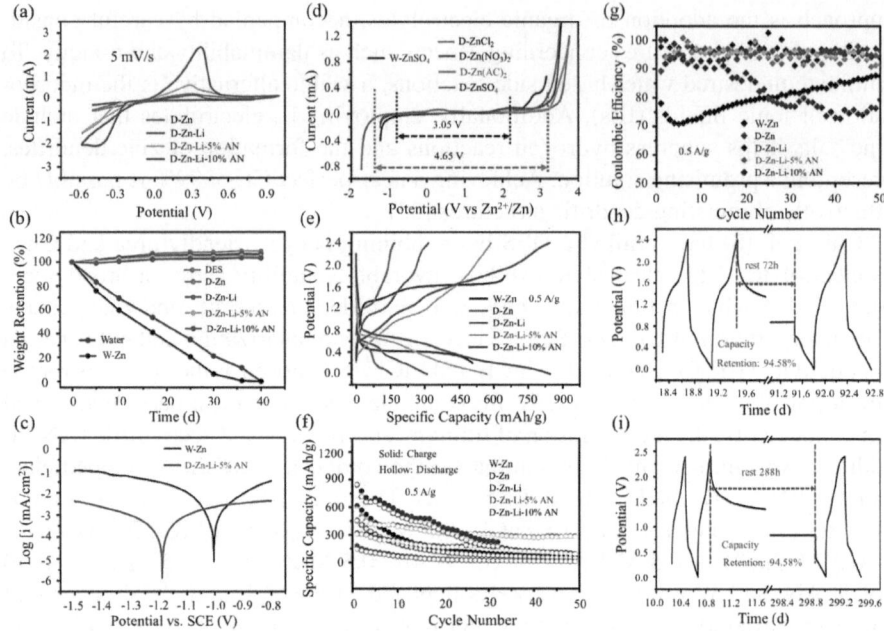

Fig. 3.15 a CV profiles depicting the processes of zinc stripping and plating. **b** Recorded weight preservation of various electrolytes exposed to ambient conditions. **c** Polarization profiles for zinc anode in the electrolytes composed of W–Zn and D–Zn–Li-5% AN. **d** ESPW was observed in various electrolytes for the engineered batteries. **e–g** GCD analysis, assessing cyclic stability, and determining CE in Zn–S batteries with varying electrolytes. **h, i** Self-discharge profiles of the battery employing D–Zn–Li-5% AN electrolyte, observed at **h** after 3 and **i** 12 days. Reproduced with permission from [8]

to 4.65 V. This observed expansion diminishes slightly upon utilizing D–Zn(Ac)$_2$, D-ZnSO$_4$, and D-Zn(NO$_3$)$_2$. In Fig. 3.15e, the graphical representations show the GCD curves under a current density of 5 A/g. It is evident that by introducing lithium salt alongside 10 vol.% AN and 5 vol.% AN into the D–Zn electrolyte, there is a notable increase in the recorded discharge capacity. Specially, the discharge capacity rises from 153.4 mAh/g to 846.1 mAh/g and 569.3 mAh/g, respectively. The contrast in measured discharge capacity between D–Zn–Li-10% AN and W–Zn highlights a considerable 67.1% superiority of D–Zn–Li-10% AN over W–Zn. Figure 3.15f, g show the endurance of capacity and CE across numerous charge and discharge cycles, providing insights into the batteries' electrochemical performance. The analysis of the collected data reveals that among the various organic and aqueous electrolytes tested, the Zn–S battery containing D–Zn–Li-5% AN electrolyte demonstrates the highest efficiency. After fifty cycles at a current density of 0.5 A/g, it maintains a remarkable 95% retention of average CE and retains almost half of its initial capacity. This suggests that D–Zn–Li-5% AN could be regarded as the top-performing electrolyte for Zn–S batteries. Observing Fig. 3.15h, i, self-discharge experiments were carried out on the battery equipped with D–Zn–Li-5% AN electrolyte. After being fully charged and left for 72 h, the Zn–S battery retained an impressive 94.58% of its initial capacity, showcasing its remarkable electrochemical prowess. By extending the duration of the fully charged Zn–S battery's idle period to 288 h, the battery's capacity retention experienced a decline to 68.58%. This resistance against self-discharge is credited to the suppression of undesirable by-reactions and dendrite formation on the anode, alongside the electrolyte's high ion diffusivity provided by D–Zn–Li-5% AN.

Cui et al. [8] introduced an electrolyte design optimized for cost-effectiveness and efficiency in Zn–S conversion batteries. They applied ChCl/urea DES as the principal electrolyte and incorporated AN as an auxiliary co-solvent, as depicted in Fig. 3.15c, d. This novel strategy effectively tackled the common drawbacks often found in conventional aqueous ZIBs. These issues encompassed challenges regarding the limited cathode-specific capacity, the formation of undesired by-products, and the occurrence of dendrite formation (Fig. 3.15e). DES proved highly soluble with diverse zinc salts and displayed a broad ESPW, successfully addressing these challenges [37–42]. Improving the DES-based electrolyte required the integration of Li^+ ions and AN, which consequently improved the ionic conductivity and advanced the transport kinetics of Zn^{2+}. CV analyses brought to light that the D–Zn–Li-10% AN electrolyte outperformed the others in terms of rapid Zn^{2+} transport kinetics and superior reversibility. Consequently, this electrolyte setup displayed a remarkable discharge capacity of 846 mAh/g_S and an impressive energy density of 258.62 Wh/kg. Notably, the battery exhibited outstanding resistance to self-discharge, attributed to minimal side reactions and the non-existence of the shuttle effect within the DES. Additionally, the symmetric zinc cell showcased zinc stripping and plating processes free from dendrite formation, leading to an impressively extended lifespan exceeding 3920 h, as depicted in Fig. 3.15f. This lifespan surpassed that achieved with conventional aqueous electrolytes by a remarkable factor of 20.6.

Concluding our investigations and delving into the realm of electrolytes, we summarize the progresses achieved in diverse types of zinc-ion electrolytes. These encompass aqueous electrolytes, both without and with supplementary additives known as redox mediators, hydrogel, and non-polar (organic) electrolytes. The efficacy of ZIBs electrochemical performance is primarily influenced by several parameters including the constituents of the electrolyte, the concentration of zinc salt, and the involvement of redox mediators. Despite considerable efforts in developing and accessing ZIBs containing aqueous electrolytes encompassing $ZnSO_4$, $Zn(CF_3SO_3)_2$, and $Zn(CH_3COO)_2$, they are hindered by intrinsic limitations. These limitations involve a restricted ESPW, pronounced polarization, the generation of irreversible side-products such as SO_4^{2-} and H_2S, and the emergence of dendritic formations on the electrodes. These limitations contribute to subpar reversibility, elevated potential hysteresis, reduced CE, insufficient thermal stability, and diminished load-bearing capacity, thereby limiting their extensive applicability. In response to the intrinsic constraints concerning the ESPW, the generation of irreversible side-products, challenges with reversibility, and elevated potential hysteresis, supplementary additive known as redox mediators, which are well-suited to the electrodes, encompassing TU, TUI, and I_2, have been integrated as a remedy. Researchers have studied the electrochemical performance of ZIBs, investigating a range of functional additives and their concentrations. They have focused on identifying reactions that help alleviate any associated drawbacks. The inclusion of redox mediators improves both reversible capacity and energy storage of the batteries through widening the ESPW, mitigating the potential hysteresis, and impeding the generation of irreversible by-products. As an illustration, the inclusion of 0.05 wt.% of I_2 into a $1 M Zn(CH_3COO)_2$ electrolyte leads to enhancements in both reversible capacity and potential hysteresis. Specifically, reversible capacity experiences an increase from 685 mAh/g to 1105 mAh/g, while potential hysteresis decreases from 0.9 V to 0.72 V [36]. The inclusion of 2 g/L of TU into $ZnSO_4$ electrolyte results in a decline in the potential hysteresis from 0.81 V to 0.46 V, accompanied by a rise in the diffusion coefficient within the electrolyte. Furthermore, while a notable degradation was evident after 20 cycles in the Zn–S battery due to the generation of SO_4^{2-} by-product, the inclusion of TU leads to only a negligible decline in capacity, amounting to merely 0.11% per cycle following 300 cycles [7]. Similar enhancements have been attained through the inclusion of 1 wt.% TUI into a $3MZnSO_4$ electrolyte. As a result of the inclusion of this component, the discharge capacity experiences a boost from 216 mAh/g to 480 mAh/g. Additionally, the potential hysteresis reduces from 1.03V to 0.72 V for the first cycle and from 0.89V to 0.5 V for the second cycle. Moreover, the inclusion of a redox mediator contributes to improved stability and reversibility [3].

Although diverse improvements can be attained by the inclusion of suitable redox mediators into aqueous electrolytes, certain intrinsic limitations of ZIBs employing aqueous electrolytes remained unaddressed. Organic or non-polar electrolytes encompassing ZnS, zinc chloride, zinc nitrate, and zinc acetate can hold the capability to furnish Zn–S batteries with prolonged longevity while offering enhanced reversible capacity, energy storage, and ionic conductivity. The elimination of irreversible side-products derived from water and the prevention of dendritic structure generation on the zinc anode contribute to a significant improvement in the delivered CE, reaching close to 100%. Another beneficial aspect provided by Zn–S batteries employing non-polar electrolytes is the widening of ESPW. This is due to the lack of water within non-polar electrolytes, which hinders the occurrence of oxygen and hydrogen evolution reactions. As an illustration, substituting an electrolyte that includes $ZnCl_2$ with one incorporating $ZnSO_4$ elevates the ESPW from 3.05 V to 4.65 V. Additionally, excluding water enhances the thermal resilience of the batteries by mitigating issues associated with water evaporation. Yet, the safety apprehensions regarding non-polar ZIBS, including the issues of flammability and toxicity, have restricted their widespread implementation.

In contemporary times, the progress in ductile and wearable electronic devices has sparked the interest of researchers in engineering Zn–S batteries with superior performance, a lightweight design, thermal durability, ductility, and capability to endure diverse loads. Considering this, gel electrolytes and their versatile counterparts, including zinc acetate, have been implemented. A quasi-solid-state battery, comprising of zinc acetate, EG, and a redox mediator containing 0.1 wt.% I_2, exhibits commendable electrochemical and mechanical proficiency across temperature ranging from − 25 to + 25 °C. While achieving a discharge capacity of 667 mAh/g_{CNF-S} and an energy density of 283 Wh/kg_{CNF-S}, the battery experiences a 60% capacity degradation following 500 cycles [33].

Despite the considerable strides undertaken thus far, as outlined briefly in Table 3.1, there exist numerous hurdles that warrant special attention. These hurdles encompass the elimination of water-induced irreversible side-products and the generation of dendritic structures, aiming to mitigate capacity degradation and ultimately improve reversibility. Additionally, there is a requirement to enhance the battery's ability to withstand varying temperatures and loads, while concurrently decreasing its overall weight. Another essential hurdle within the electrolyte of Zn–S batteries lies in the onset of the HER. This reaction tends to be activated under circumstances, like during over-discharge or when the durability of the electrolyte is jeopardized. The outcome of this reaction is the production of hydrogen gas, which exhibits inefficiency and is also a potential safety hazard [29, 43]. The occurrence of the HER is influenced by the chemical composition, pH level, and potential window of the electrolyte.

3.3 Cathode Configuration

The progression of Zn–S battery technology encounters obstacles with the most prominent being the sluggish nature of the reactions occurring between zinc and sulfur. The chief reason behind their underwhelming performance stems from the inherent inadequacy of the redox potential associated with the Zn–S reaction. The transformation from sulfur to ZnS encompasses a two-electron electrode reaction, characterized by its inherent fragility and limited efficiency. Cai et al. [30] undertook a recent investigation aimed at tackling this hurdle, leading to a considerable 380% surge in energy density, surpassing the previously recorded maximum values by other researchers [13]. This highlights the significance of fine-tuning the reaction conditions of the sulfur cathode.

Significant research endeavors have notably propelled the domain of Zn–S batteries forward, with particular emphasis on advancements in cathode design, electrolyte chemistry, and overall system architecture. The introduction of carbon additives into sulfur cathodes has revolutionized the realm of Zn–S batteries, provided vital structural reinforcement, and significantly improved electrochemical efficiency. Innovative cathode composites and optimized electrolyte formulations, denoted as "cocktail-optimized", have been investigated to enhance energy density, capacity, and longevity, frequently yielding remarkable outcomes. Unconventional and innovative methodologies, encompassing the integration of CNF, the engineering of HCS architectures, employing cathode with yolk-shell architecture, and utilizing distinctive electrolyte solvents, have been adopted to tackle the hurdles pertaining to sulfur redox kinetics and battery deterioration. The combined progress made in these areas

Table 3.1 A comparative analysis of the electrochemical efficiency of aqueous, non-polar, and multifunctional electrolytes. Reproduced with permission from [11]

Electrolyte	Cathode material	Cathode active material loading (mg/cm²) [2]	Anode composition	Voltage window	Redox mediator	Energy density	Max capacity	Current density	Cycle	Capacity retention	Refs.
3M ZnSO₄	ZnS@CF	1.5–3.0	Zinc foil	0.1–1.4 V	Iodinated Thiourea	832 Wh/kgS 274 Wh/kgZnS	1410 mAh/gS (465 mAh/gZnS)	0.1 A/g	300	65% at 2 A/g	[3]
1 M Zn(CH₃COO)₂	S@CNT	3.0–5.0	Zn foil	0.05–1.6 V	0.05 wt% I₂	502 Wh/kgS	1105 mAh/gS	0.1 A/g	225	65% at 2 A/g	[14]
1 M ZnCl₂	KB-S composite	3.9–8.3	Zn foil	0.1–0.95 V	-	1083.3 Wh/kgS	1668 mAh/gS	50 mA/g	-	80.1%	[13]
3M Zn(OTF)₂	S@CMK-3	1.0–2.0	Zn foil	0.05–1.75 V	I₂	378 Wh/kgS	788 mAh/gS	0.2 A/g	100	65% at 1 A/g	[15]
2M Zn(OTF)₂	S@HCS	1.2–1.6	Zn foil	0.1–1.6 V	tetraglyme + I₂	570 Wh/kg	1140 mAh/g	0.5 A/g	600	70% at 4 A/g	[44]
2M Zn(OTF)₂	S@HCS	1.2–1.6	Zn foil	0.1–1.6 V	tetraglyme + I₂	460 Wh/kg	937 mAh/g	1.7 A/g	130	80% at 1.7 A/g	[44]
3 M Zn(OTF)₂	HMCs-3@S	1.0–2.0	Zn foil	-	-	307 Wh/kg	768 mAh/g	0.1	100	94% at 1 A/g	[45]
2 M Zn(OTF)₂	S@NPC	3.0	Zn foil	0.1–1.6 V	ethylene glycol + ZnI₂	730 Wh/kg	1435 mAh/g	0.1	250	70% at 3 A/g	[18]
2 M ZnSO₄	S@KB	1.0	Zn foil	0.2–1.75 V	2 g/L Thiourea	803.3 Wh/kgS	1740 mAh/gS	0.5 A/g	300	51% at 5 A/g	[7]

(continued)

Table 3.1 (continued)

Electrolyte	Cathode material	Cathode active material loading (mg/cm² [2])	Anode composition	Voltage window	Redox mediator	Energy density	Max capacity	Current density	Cycle	Capacity retention	Refs.
2MZn(CH$_3$COO)$_2$	S@CNF	3.0–7.0	Zn foil	0.1–1.6 V	0.1 wt.% I$_2$ and ethylene glycol +	283 Wh/kgCNF–S 149 Wh/kgZnS	667 mAh/gCNF–S	0.5 A/g	300	40% at 1A/g	[33]
1 M ZnCl$_2$ + DES	S@KB	1.0–1.5	Zn foil	– 0.6–1.0 V	LiCl$_2$ + AN	259 Wh/kgS	846 mAh/gS	0.5 A/g	400	38% at 1 A/g	[8]
Aqueous alkali/acid hybrid three electrolyte configuration	S/C cathode	–	Zn foil	1.55–2.15 V	–	3800 Wh/kgS	2250 mAh/g	10 A/g 1 A/g	400	86.5% at 5 A/g	[30]
0.5MNaOH (anolyte)and 0.1 M Na$_2$S$_4$ + 0.1 M NaOH	CoS@SS	–	Zn foil	0.3–0.9 V	–	586 Wh/kgNa$_2$S$_4$	966 mAh/gNa$_2$S$_4$	0.5 mA/cm^2	50	85%	[23]
1 M Zn(TFSI)$_2$ + 21 M LiTFSI	PLSD cathode	4.0	Zn foil	0.1–2.3 V	–	724 Wh/kgcathode	1148 mAh/gS	0.5 A/g	750	Capacity retention of 204 mAh/gS after 1600 cycles at 1 A/g	[29]
1 M Zn(CH$_3$COO)$_2$	S@CNTs	1.5–2.5	Zn foil	0.1–1.6	PEG-400	892 Wh/kgS	1116	0.1 A/g	300	81% at 1A/g	[46]
2 M ZnSO$_4$	S@Fe-PANi	1.8	Zn foil	0.1–1.6	–	720 Wh/kgS	1205	0.2 A/g	200	54% at 0.5 A/g	[32]
ZnSO$_4$ + PVA	S@FeNC/NC/CC	2.0–2.6	Zn foil	0.1–1.6	–	777 Wh/kgS	1143	0.2 A/g	300	58% at 0.5 A/g	[43]
2 M ZnSO$_4$	OS	2.0–3.0	Zn@2DZS	0.1–1.6	I$_2$	866 Wh/kgS	1295	1 A/g	5000	94% at 1A/g	[31]
2 M ZnSO$_4$	S@CMK-3	2.0–2.5	pZn/In		ZnI$_2$		1630	0.1 C	300	51% at 2 C	[47]

highlights the promising outlook for Zn–S batteries, suggesting their potential to achieve significant capacity, sustained stability, and considerable energy density.

CNFs provide augmented structural support, whereas HCS offers dual advantages by suppressing the formation of dendrite structures and enhancing ion transfer. The progress achieved through these innovations elevates the effectiveness and longevity of batteries, rendering them suitable for a multitude of applications. Expanding the implementation of these pioneering battery technologies encounters numerous hurdles. The intricacies of manufacturing and the time required to bring carbon spheres or fibers to market can impact their quality and competitiveness in the market. Moreover, maintaining uniform performance levels across various manufacturing methods is imperative. This is particularly crucial for cathodes utilizing HCS architectures infused with sulfur, where the precise sizing and uniformity of sulfur and carbon particles become paramount. Any inconsistencies in the mentioned factors may lead to adverse effects on fundamental battery performance, encompassing energy density, charging efficiency, and longevity. Hence, meticulous consideration of these subtleties will be essential as we move from experimental models in the laboratory to full-fledged commercial offerings. Further exploration and advancement could yield remedies to alleviate these hurdles, ultimately clearing the path for the extensive utilization of these inventive cathode configurations. The inclusion of such additives, which hold the promise of improving electrical properties and bolstering structural integrity, marks a promising frontier in the ongoing endeavor to develop batteries that exhibit heightened efficiency and extended longevity. Research findings indicate promising outcomes from incorporating carbon fibers or alternative metals into sulfur cathodes. Generally, this section delves into the hurdles confronting Zn–S battery cathodes, which impede their real-world applications. Moreover, we will investigate the diverse array of cathode designs that have been developed thus far to address these hurdles.

3.3.1 Cathode Materials and Configurations

Under typical ambient conditions, sulfur materials demonstrate an extraordinary minimal level of conductivity, measuring at 5.0×10^{-30} S/cm. In response to the issues surrounding reaction kinetics and conductivity, manufacturers often design composite cathodes by either integrating sulfur with conductive materials and polymer binders or by embedding sulfur within a matrix of conductive hosts. Maximizing battery performance relies heavily on the meticulous development and selection of cathode materials, considering the intricacies of storage mechanisms. Although there have been recent strides in optimizing strategies for cathode materials in Zn–S batteries, there persist several unresolved challenges. Zn–S batteries face challenges that hinder their practical implementation due the various factors, including:

3.3.1.1 Sluggish Cathode Kinetics

Within the domain of Zn–S batteries, the phrase "Sluggish Cathode Kinetics" refers to the delayed progression of reactions or ion movement within the cathode substance of the battery. The delayed nature of this phenomenon often stems from the strong electrostatic interaction and significant steric hindrance between divalent Zn^{2+} ions and the inherent composition of cathode materials in aqueous ZIBs. Consequently, this leads to less than-optimal cycling performance and notably sluggish intercalation kinetics. Besides inadequate conductivity, the produced ZnS compound exhibits a diminished solubility product (Ksp), thereby raising resistance within the cell. Moreover, the sulfur cathode's poor wetting properties with the aqueous electrolyte hinder the transportation of Zn^{2+} ions, leading to considerable polarization. In the pursuit of the kinetics of reactions, redox mediators are commonly introduced into either the electrolyte or the host cathode. Furthermore, addressing the sluggishness in cathode kinetics also entails the optimization of the structure of the sulfur cathode. Advancements in cathode architecture, such as the fabrication of materials with enhanced porosity or nanostructured features, can offer additional routes for ion migration and promote accelerated electrochemical reactions [23, 43].

3.3.1.2 Conductivity Hurdles in Sulfur and Zinc Sulfide

Due to inadequate conductivity exhibited by sulfur, employed as the cathode material, and ZnS produced as the discharge product, the cathode encounters substantial internal resistance and polarization throughout the charging and discharging processes within the cell. Hence, adjustments to the electrode are imperative to boost the conductivity of both S and ZnS. One prevalent technique involves integrating conductive carbon to serve as a scaffold for sulfur, enclosing it, and thereby activating its electrochemical properties. Further elaboration concerning alternations to the cathode in a Zn–S battery is outlined in the following section.

3.3.1.3 Alternations in Volume Occur at the Cathode

The marked disparity in density between S, standing at 2.3g cm^{-3}, and ZnS, standing at 4.09 g cm^{-3}, results in a significant volume shift (50.3%) during the cycling process. Consequently, the electrode undergoes pulverization, leading to the formation of isolated areas where the active constituents lose their electrical connectivity with the conductive matrix, thereby rendering them inactive. Dealing with this issue requires strategies such as enveloping sulfur within a porous conductive carbon framework or engineering electrode materials with the flexibility to adapt to volume alterations, thereby ensuring robust mechanical resilience and prolonging cycle longevity.

3.3.1.4 Durability of the ZnS Deposits

The pH level of the electrolyte significantly impacts the durability of sulfides. ZnS, produced throughout the discharging process, exhibits instability in acidic environments, leading to a disproportional reaction that restricts the utilization of acidic electrolytes in Zn–S batteries. Approaches like integration of additives or employing concentrated electrolytes serve to balance the pH of the electrolyte, effectively mitigating this challenge.

3.3.1.5 Creation of Sulfate Ions

When sulfur in the cathode encounters dissolved O_2 and ZnS interacts with H_2O throughout the discharging process, it produces an irreversible by-product, sulfate ions. This outcome compromises electrode reversibility and leads to the depletion of active sulfur material. By refining the synthesis methodology to mitigate the direct contact of sulfur with the surface and effectively eliminating dissolved O_2, the cycling efficiency of the cathode can be enhanced.

In response to the constraints, researchers have introduced two distinct types of cathodes: one designed to address the sulfur utilization problem, while the other deals with stabilizing the cathode structure. The investigations led by Zhao et al. and Liu et al. center around optimizing sulfur utilization in Li–S batteries. Their research delved into the use of redox mediators to facilitate the conversion of lithium polysulfides and alleviate the shuttle phenomenon [15, 31]. In the domain of Zn–S batteries, investigations by Gross and Manthiram, alongside those of Guo et al., delved into methods to improve sulfur conversion efficiency and overall battery performance. Guo et al.'s study introduced a novel hybrid electrolyte design aimed at achieving these goals [18, 23]. In the realm of preserving cathode structural integrity, Zhang et al. introduced a yolk-shell-structured sulfur cathode tailored for Zn–S batteries. Their focus laid in mitigating cathode deterioration [32]. Similarly, Amiri et al. delved into the creation of a quasi-solid-state Zn–S battery, emphasizing the importance of upholding cathode stability. They utilized an innovative coating methodology, referred to as electrostatic spray, on activated carbon nanofibers as part of their approach [33]. Collectively, these investigations signified substantial progress in addressing the hurdles associated with sulfur utilization and maintaining cathode stability within sulfur-based battery technologies. These initiatives encompass developments in both carbon/sulfur composite cathodes as well as hybrid cathode designs.

3.3.2 Carbon/Sulfur Composite Cathodes

A multitude of strategies have been investigated to advance the development of batteries that not only deliver enhanced performance but also uphold structural durability. Referenced studies illustrate diverse tactics for achieving this formidable objective [3, 9]. A highly promising direction involves the encasement of sulfur cathodes within one-dimensional carbon fibers (1D CNFs). The techniques that were employed to produce these 1D cathodes exhibit diversity across various research endeavors, highlighting the strength and flexibility inherent in this method. The use of CNFs as a host material for sulfur encapsulation is motivated by the intention to exploit the superior electrical conductivity of CNFs to address the inherent low electrical conductivity of sulfur. Sulfur, heralded for its potential in energy storage applications, faces hurdles due to its inherent limitation in electrical conductivity. To overcome this constraint, sulfur is enclosed within the matrix of CNFs, thus capitalizing on the exceptional electrical conductivity characteristics of CNFs. The hybridization strategy seeks to synergize the advantages of both sulfur and CNFs, augmenting the composite material's overall electrical conductivity to achieve superior performance in energy storage applications. Within the existing studies, investigations employing sulfur cathodes incorporated within CNFs (commonly denoted as S@CNF) have yielded notable findings.

The multifunctional nature of CNFs is evident. Initially, they serve as a protective barrier, shielding the sulfur cathodes from environmental elements that could potentially diminish their performance. This protective feature has the potential to enhance the battery's longevity and sustain its efficacy over time. Another aspect is that the CNF layer can provide structural strength to the cathode, essential for maintaining the battery's robustness across a range of operating conditions. For applications where mechanical durability and resilience are equally paramount to electrical performance, this structural assistance proves indispensable. It should be emphasized that sulfur loading can be adjusted to meet applications or performance criteria, as evidenced by the range of preparation techniques described in different studies. These distinctions are evident in the construction of the CNF/CF cathodes, as illustrated in Fig. 3.16a, b. The visual contrasts between them are apparent in Fig. 3.16b, d. The versatility of this method further highlights its potential for extensive utilization in advanced batteries of the future, which require both outstanding performance and robust structural durability.

Amiri et al. [33] introduced a technique for fabricating S@CNF tailored for Zn–S batteries. The manufacturing process is initiated with electrospinning polyacrylonitrile-polymethylmethacrylate (PAN-PMMA) polymers to produce porous CNFs, with decomposing PMMA serving as the sacrificial material. Following this step, the fibers underwent activation in a nitrogen-rich environment after being soaked in a potent alkaline KOH solution. This process prompts the creation of external micro- and mesopores, as illustrated in Fig. 3.16a. Moreover, this method induced the incorporation of oxygen-containing functional groups, thereby enhancing the chemical reactivity of the electrode. Following this, sulfur is deposited onto the surfaces and pores of the activated fibers using an electrostatic spray technique, following precise guidelines regarding environmental conditions and the ratio of fiber to sulfur. Notably, Amiri et al. [33] illustrated that the S@CNF configuration not only upholds the structure resilience of Zn–S batteries but also maintains their electrochemical efficacy, with particular attributes observable in Fig. 3.16b. By

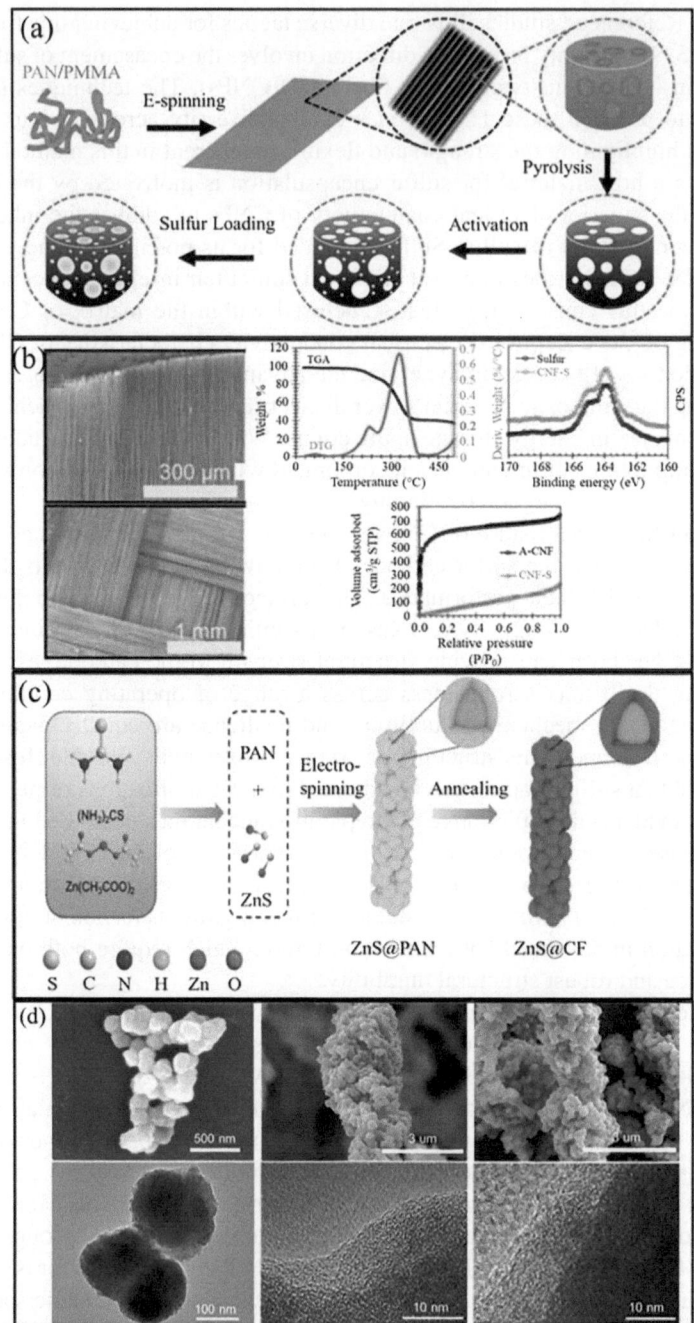

Fig. 3.16 **a** Amiri et al. [33]'s methodology for fabricating CNT cathodes, **b** scanning electron microscopy (SEM) imagery depicting the cathode alongside its corresponding chemical performance [33], **c** schematic procedure outlined by Liu et al. [3] for fabricating the CF cathode, **d** SEM and TEM visualization of the corresponding nanoparticles [3]. Reproduced with permission from [3, 33]

employing an electrostatic spray coating method, they successfully attained a sulfur loading exceeding 60 wt.%, effectively occupying more than 69% of micro- and mesoscale pores. This achievement was enabled by the extensive surface area and substantial pore volume of the activated CNFs, which were further enhanced through hot-drawing techniques. Regarding outcomes, the structural Zn–S battery integrating S@CNF alongside a freeze-resistant hydrogel electrolyte surpassed current ESSs, boasting a maximum energy density of 283 Wh/kg. The material demonstrated remarkable mechanical features, showcasing a tensile strength reaching 377 MPa and Young's modulus of 16.7 GPa. Notably, both the electrochemical and mechanical traits exhibited robust stability, even following prolonged cycling. As an instance, the electrochemical features remained consistent throughout 10,000 cycles of mechanical tensile and release cycles, whereas the mechanical resilience experienced negligible degradation following 3,000 electrochemical cycles, which is known as E-cycles. Amiri et al. [33]'s innovative method not only increased the energy density of Zn–S batteries but also established a durable and steadfast framework, representing a considerable leap forward in the domain of structural ESSs.

Liu et al. [3] combined conventional methodologies with tailored approaches to enhance Zn–S battery's performance. They commenced by synthesizing ZnS nanoparticles through the dissolution of anhydrous zinc acetate and Poly (N-vinyl-2-pyrrolidone) (PVP) in EG, followed by introducing TU into the solution, as depicted in Fig. 3.16c. Upon heating, the ZnS nanoparticles underwent extraction via centrifugation and subsequent rinsing steps, as depicted in Fig. 3.16d. Concurrently, two different cathode configurations were devised: one featuring Zn–S with carbon and the other incorporating Zn–S with carbon fiber, identified as ZnS@C and ZnS@CF, respectively.

In the case of ZnS@C, a mixture comprising the synthesized ZnS nanoparticles and PAN dissolved in N, N-dimethylformamide (DMF), referred to as "Solution A", underwent a gradual heating process. ZnS@CF was produced via an electrospinning technique employing "Solution A", followed by an additional heating step to eliminate any remaining DMF. For the anode, referred to as Zn@PAN, PAN was dissolved in DMF, and a ZnS compound was added. The resulting mixture was subsequently applied onto a pre-treated zinc foil, finalizing the anode fabrication. These thorough endeavors led to a notably improved battery performance. When compared with a precisely adjusted $ZnSO_4$ electrolyte, the ZnS@CF cathode showcased a considerable specific capacity, reaching 465 mAh/g_{Zn-S} (equivalent to 1410 mAh/g_S) and an energy density of 274 Wh/g_{Zn-S} (equivalent to 832 Wh/kg_{Zn-S}) at the current density of 0.1 A/g. Additionally, the battery exhibited long-term stability, maintaining a capacity exceeding 226 mAh/g even after undergoing 300 cycles at a current density of 2 A/g. Study [3] serves as a prime example of how intricate material synthesis and refined architectural designs significantly impact the comprehensive performance parameters of Zn–S batteries. By integrating carbon fiber nanoparticles, the study achieved substantial improvements in both capacity and energy density. It amalgamates inventive preparation methodologies with rigorous performance assessment, underscoring the promising potential of such batteries for applications requiring high-capacity energy density.

The incorporation of carbon additives into sulfur cathodes for Zn–S batteries presents a diverse range of advantages, improving the comprehensive efficacy and resilience of ESSs. Carbonaceous substances, spanning from nanofibers to conductive carbon black, not only encase the sulfur cathodes but also provide organized spaces to accommodate sulfur elements. Furthermore, they can act as reinforcing films or textiles to facilitate superior bonding of the cathode [8, 12, 18, 29, 32, 44].

Cui et al. [8] developed an advanced cathode for Zn–S batteries by combining conductive carbon black and sulfur at a weight ratio of 2:1. This blend was subjected to thermal processing at 150 °C for a period of 12 h in an autoclave, leading to the development of a unique cathode material termed sulfur embedded in carbon (S@C). Coin-style zinc batteries typically incorporate a blend of active materials, KB, and polyvinylidene fluoride (PVDF) at a weight ratio of 8:1:1. This mixture is subsequently converted into a slurry using methyl-2-pyrrolidinone (NMP) solvent and deposited into carbon cloth or titanium foil as substrates. The electrode disks produced possess a sulfur mass loading ranging from 1.0 to 1.5 mg/cm^2. Utilizing discoveries from their research [8], the application of a ChCl/urea-based DES electrolyte brought about a notable enhancement in the performance of the Zn–S battery. The DES electrolyte, enriched by incorporating AN solvent, demonstrated impressive ionic conductivity, and successfully mitigated side reactions and the emergence of zinc dendrite structures. This led to the zinc anode sustaining consistent cycling for more than 3,920 h without encountering any instances of short circuits. The Zn–S battery exhibited an outstanding discharge capacity, reaching 846 mAh/g$_S$, and an exceptional energy density of 258.62 Wh/kg. Moreover, its robustness against self-discharge became apparent, thanks to the dependable performance of the zinc anode and the frequent manifestation of "shuttle issues". More precisely, the battery preserved 94.58% of its capacity after 72 h of rest, and 68.58% after 288 h of rest. The multifunctionality of the DES electrolyte implies its suitability for deployment in diverse ZIB systems. With the integration of cutting-edge cathode architecture and the advanced DES electrolyte, Zn–S batteries are on the brink of experiencing revolutionary enhancements in discharge capacity, energy density, and overall durability.

Guo et al. [18] engineered a distinctive hybrid aqueous electrolyte designed specifically for Zn–S batteries. This novel formulation boasts a wide ESPW compared to conventional aqueous electrolytes. The hybrid formulation serves to mitigate sulfur side reactions occurring at the cathode while also preventing the formation of zinc dendrite structures on the surface of the zinc metal anode. The introduction of EG into the electrolyte facilitates hydrogen bonding with unbound water molecules, effectively reducing undesirable side reactions involving these molecules. Additionally, EG promotes the speed of zinc-ion movement by facilitating its coordination with zinc ions. Impressively, the Zn–S battery utilizing nonporous carbon, known as ZnS@NPC, demonstrated an exceptional capacity of 1,435 mAh/g and an outstanding energy density of 730 Wh/kg, when operated at a current density of 0.1 A/g. Even when subjected to a high current density of 3 A/g, the battery consistently sustains a capacity surpassing 300 mAh/g throughout 250 cycles. Uniquely, rather than following the conventional single-step conversion from S to ZnS in an aqueous electrolyte, this hybrid approach utilizes a multi-step conversion procedure involving both S and ZnS. This innovative employment of a cost-effective hybrid electrolyte provides a remedy to the issues associated with sulfur side reactions encountered in different metal-sulfur battery configurations. It introduces a new dimension to Zn–S electrochemistry, offering a promising direction for future research in Zn–S battery technology.

3.3.3 Hybrid Cathodes

Zhao et al. [29] innovated a methodology for improving Zn–S batteries, drawing inspiration from the protective strategies of the tardigrade. Employing carbon cloth as a substrate stemmed from a purposeful decision due to its exceptional stability in facilitating the transfer of ions. The formation of chemical bonding between the carbon cloth and a specialized copolymer known as poly(Li_2S_6-r-DIB) or PLSD obviates the requirements for supplementary binders or conductive additives. Furthermore, an IL was incorporated into the design to augment the transport of Zn^{2+} ions. To produce Zn/IL-PLSD and Zn/LF-PLSD batteries, the initial step involves the application of the PLSD slurry onto carbon cloth using a doctor blade, followed by sintering at 150 °C for 20-min period of time, as depicted in Fig. 3.17. As it was mentioned, the fabrication procedure of the cathode does not necessitate any binders or conductive additives, as the copolymer and the carbon cloth form a chemical bonding, allowing for an impressively high PLSD load of 4 mg/cm^2. Following this step, the PLSD cathode was coated with 4-(3-butyl-1-imidazolio)-1-butanesulfonate IL and subsequently left to air dry at a temperature of 50 °C for half an hour, yielding what is referred to as the IL-PLSD cathode. The selected ILs satisfy two conditions: they retain liquidity at ambient temperature, thereby aiding the mobility of Zn^{2+}, and they feature sizable anions such as $CF_3SO_3^-$ which act as conduits for Zn^{2+} transportation. Additionally, other ILs that meet these specifications could also be considered as alternatives.

Figure 3.17a illustrates how this design led to the development of a Zn/LF-PLSD battery that yielded impressive outcomes. The battery exhibited outstanding performance, delivering a remarkable capacity of 1,148 mAh/g and a considerable energy density of 724.7 Wh/kg cathode at current density of 0.3 A/g. Despite enduring 700 cycles at current density of 1 A/g, it sustained a capacity exceeding 235 mAh/g, highlighting the enduring stability of the sulfur electrode created. The investigation into mechanisms unveiled the electrochemical processes accountable for these results. Zinc predominantly reduces S_6^{2-} to S^{2-} throughout the discharging process, followed by its oxidation to generate long-chain $Zn_xLi_yS_{3-6}$ throughout the charging process. Through the implementation of a highly concentrated electrolyte, the team fine-tuned the system, resulting in an extended operational lifespan of 1,600 cycles at the current density of 1 A/g, while retaining a capacity of 204 mAh/g. The introduction of a highly concentrated electrolyte also resulted in a reduction in the solubility of polysulfides, offering an additional deterrent against the shuttle effect. Enhanced

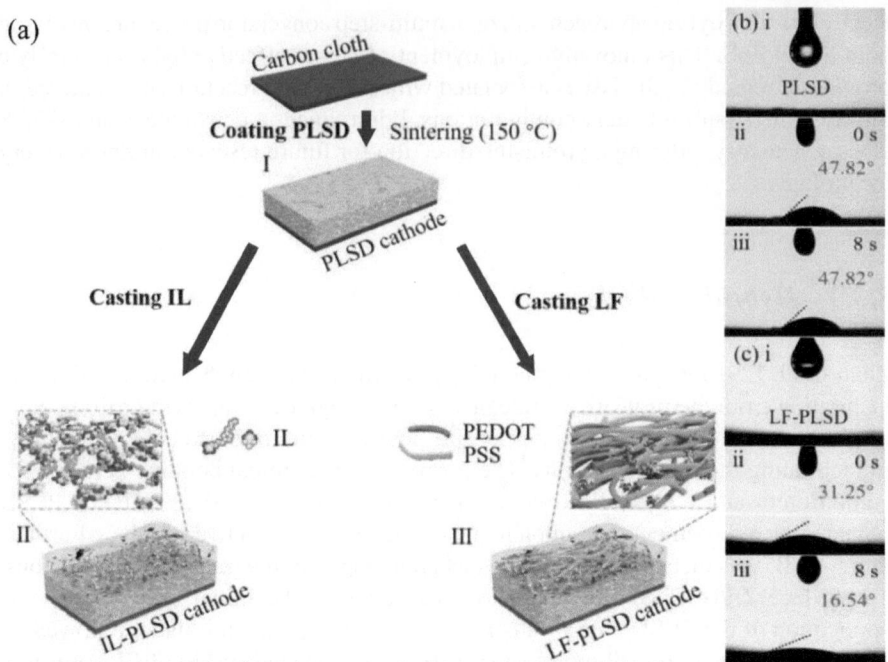

Fig. 3.17 **a** Illustration demonstrating the fabrication procedure of the PLSD cathode, alongside, **b, c** images showcasing the dynamic contact angles observed on the cathode surface when treated with PLSD and LF-PLSD, respectively. Reproduced with permission from [29]

hydrophobic features contribute to this effect, as depicted in Fig. 3.17b, c. The implementation of the liquid film (LF) methodology introduced a paradigm of sulfur-based batteries employing Zn/S chemistry. This methodology not only resolves the persistent cathode/electrolyte incompatibility challenge encountered in numerous metal-sulfur batteries but also unveils promising prospects for ZIBs.

Yang et al. [44] directed their attention to HCS as a stabilizing scaffold, implementing an exhaustive methodology that encompassed the utilization of various compounds, including tetraethyl orthosilicate (TEOS). Yet, their most striking finding was the development of an optimized electrolyte "cocktail-optimized" for aqueous Zn–S systems. This innovation significantly boosted the reversibility of both the sulfur cathode and the zinc anode. After conducting both physical and chemical examinations, the research delved into the effects of the "cocktail-optimized" electrolyte. It was discovered that the G_4 co-solvent present within the electrolytes synergistically interacted with I_2 molecules, triggering the activation of the polar I_3/I_4 catalyst couple while concurrently impeding water access to the sulfur cathode. This resulted in the development of an organic–inorganic hybrid SEI on the surface of the zinc anode.

This interface efficiently curtailed adverse side reactions such as the production of sulfate triggered by water, hydrogen evolution, and the corrosion of the electrode.

To produce a sulfur cathode using a conventional method, TEOS was blended with a solution comprising ethanol, deionized water, and ammonium hydroxide. Following agitation, resorcinol and formaldehyde were introduced into the solution, which was subsequently stirred at ambient temperature for an extended duration. Afterward, the resultant mixture underwent multiple purification steps using deionized water and ethanol. The obtained compound, denoted as $SiO_2@RF$, was subsequently subjected to drying and annealing processes to transform it into $SiO_2@C$. In the process of HCS fabrication, the removal of the SiO_2 template was achieved by treating it with a NaOH solution. Subsequently, the spheres were mixed with sulfur powder, and the resulting mixture was then sealed and subjected to heat. Similar procedures were employed to load sulfur onto other materials, including AC, CNT, and ordered mesoporous carbon material (CMK-3). Upon completion, the glass tubes were unsealed, exposing an array of sulfur-impregnated materials, which included HCS/S with varying sulfur contents, AC/S, CNT/S, and CMK-3/S.

The research conducted by Zhang et al. [32] delved into the examination of cathodes with yolk-shell architecture designed for Zn–S batteries. These cathodes underwent fabrication utilizing a distinct procedure that included aniline in conjunction with either sodium ferricyanide or ammonium persulfate. Employing this procedure resulted in cathodes demonstrating impressive reversible capacity, coupled with improved power and energy densities. In the framework of the yolk-shell architecture, as illustrated in Fig. 3.18a, b, the combination of sulfur with iron cathode, denoted as S@Fe-PANI, and a $PVA - ZnSO_4$ gel electrolyte yielded noteworthy outcomes. The polymer shell, augmented with $Zn_xFe^{II/III}(CN)_6$ redox centers, served as a storage site for cations, substantially hastening the otherwise sluggish sulfur redox processes. This led to the Zn–S cells demonstrating an impressive capacity of 1,205 mAh/g and an energy density of 720 Wh/kg_{sulfur} for solid-state cells, while showing 375 Wh/kg_{sulfur} for others, as evidenced in Fig. 3.19a. Although challenged by declining performance resulting from the emergence and buildup of inert ZnS nanocrystals and the constraint of sulfur reversibility, the mock-up ductile and wearable zinc battery exhibited steadfast capacities and outstanding mechanical durability, proving its resilience when subjected to demanding testing conditions. This investigation not only achieved a notable milestone within the field but also offered invaluable perspectives into the realm of redox electrocatalysis and the adjustment of sulfur electrochemical processes.

Figure 3.19a displays the outcomes of the investigation conducted by Zhang et al. [32], highlighting their successful development of sulfur cathodes featuring yolk-shell architectures. These designed cathodes exhibited outstanding performance parameters, notably in the examination of voltage in correlation with specific capacity. Figure 3.19b provides insight into the influence of varying concentrations of Fe^{2+} or Fe^{3+} on the binding energy of the electrode. In Fig. 3.19c, d, the change in specific capacity of the S@Fe-PANI cathode relative to the current flow throughout electrochemical investigations are depicted. This exceptional improvement provides convincing proof that integrating Fe-PANI into the yolk-shell architecture of sulfur cathodes considerably boosts the current density achievable from Zn–S batteries. This

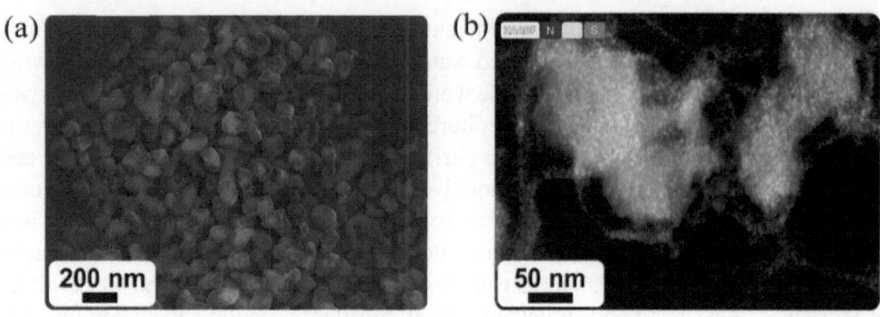

Fig. 3.18 Investigation into the structure and composition of the sulfur cathode with yolk-shell architecture: representative **a** SEM and **b** TEM micrographs of S@Fe-PANI cathode. Reproduced with permission from [32]

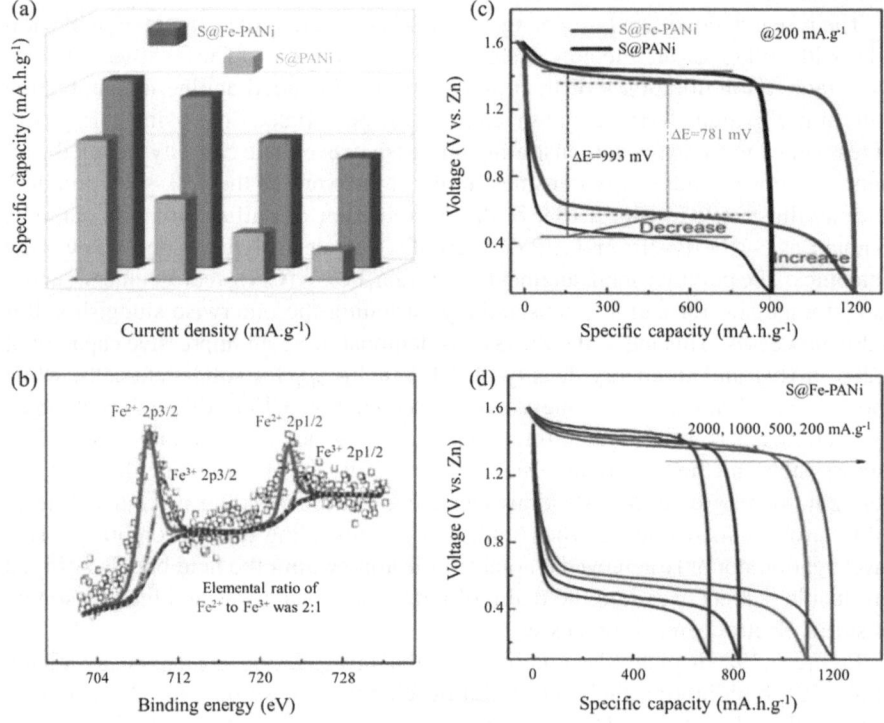

Fig. 3.19 **a** Specific capacity, derived from the charging and discharging profiles, is plotted against current density. **b** Detailed high-resolution XPS analysis of Fe 2p for the S@Fe-PANI composite. **c** Charging and discharging behaviors of S@Fe-PANI and S@PANI at 200 mA/g current density. **d** Assessment of the Zn–S coin cell's performance at different rates, utilizing a S@Fe-PANI cathode, with current densities ranging from 200 to 2,000 mA/g. Specific capacity calculated from the charge/discharge curves as a function of current density. Reproduced with permission from [32]

considerable boost in the current density makes this battery configuration particularly beneficial for a multitude of applications.

A plethora of studies have investigated a wide array of enhancers for sulfur cathodes, encompassing KB, carbon disulfide, CMK, and various other innovative compounds [7, 13, 15, 23, 31, 43]. Chang et al. [7] developed a sulfur and KB electrode, denoted as S@KB, through a melt-diffusion technique, placing significant emphasis on the structure and performance of Zn–S batteries. The production process commenced with the creation of the S@KB active material, accomplished through the blending of sulfur and KB. Following this, the blend underwent thermal treatment within a sealed quartz chamber under an atmosphere containing argon gas for a defined period. Afterward, a unique slurry composed of the S@KB active material, extra KB, and polytetrafluoroethylene (PTFE) was processed to form a thin film, yielding an electrode characterized by a considerable sulfur loading density. The effectiveness of this electrode did not solely arise from its structural layout; it also benefited significantly from the cooperative influence of TU, an electrolyte supplement. Insights into the remarkable impact of TU were gleaned from advanced ex situ Fourier Transform Infra-Red (FTIR) and XPS analyses, as depicted in Fig. 3.20a, c. Observations suggest that TU intermediates, possessing discernible positive and negative poles, engage with ZnS, resulting in the attenuation of Zn–S bonds and consequent improvement in the electrochemical behavior of the electrode. Furthermore, the carbonium ions inherent with the composition of TU displayed significant reactivity with ZnS, impeding the generation of SO_4^{2-} ions, effectively. Consequently, the electrode demonstrated not only a considerable capacity, reaching 763.7 mAh/g after the completion of 300 charging and discharging cycles at the current density of 5 A/g but also exceptional cycle stability, demonstrating a degradation rate of only 0.11% per cycle.

Luo et al. [13] utilized melt-diffusion methodology to produce sulfur and carbon composites, with a particular emphasis on integrating KB additives. Their investigation centered on the meticulousness needed in the fabrication of electrode materials, highlighting how this methodology could amplify the performance of the battery [13]. The synthesis of sulfur and carbon composites was accomplished through the employment of the melt-diffusion methodology. At first, sulfur and KB were hand-crushed together utilizing a mortar and pestle. Subsequently, the combination underwent ball milling for multiple hours. Ultimately, the obtained composite material was gathered, compacted into a pellet, enclosed in an autoclave, and subjected to heating for a predetermined time. Figure 3.20d, e illustrate XRD pattern, binding energy analysis, and SEM micrographs of these cathodes. Figure 3.20 illustrates both cathodes in their discharged state, demonstrating the influence of the electrolytes on the cathode surfaces. These observations provide a basis for comparison with the ZnS@KB cathodes developed by Chang et al. [7], showcasing the potential disparities in electrochemical performance stemming from different fabrication methodologies. Figure 3.20a–c showcase the XRD pattern, binding energy analysis, and SEM micrographs conducted by Chang. Contrasting the electrolyte composition highlights the impressive effect on both the battery's discharge capacity and the plateau voltage. With the application of a 1 M $ZnCl_2$ electrolyte, the sulfur cathode delivered

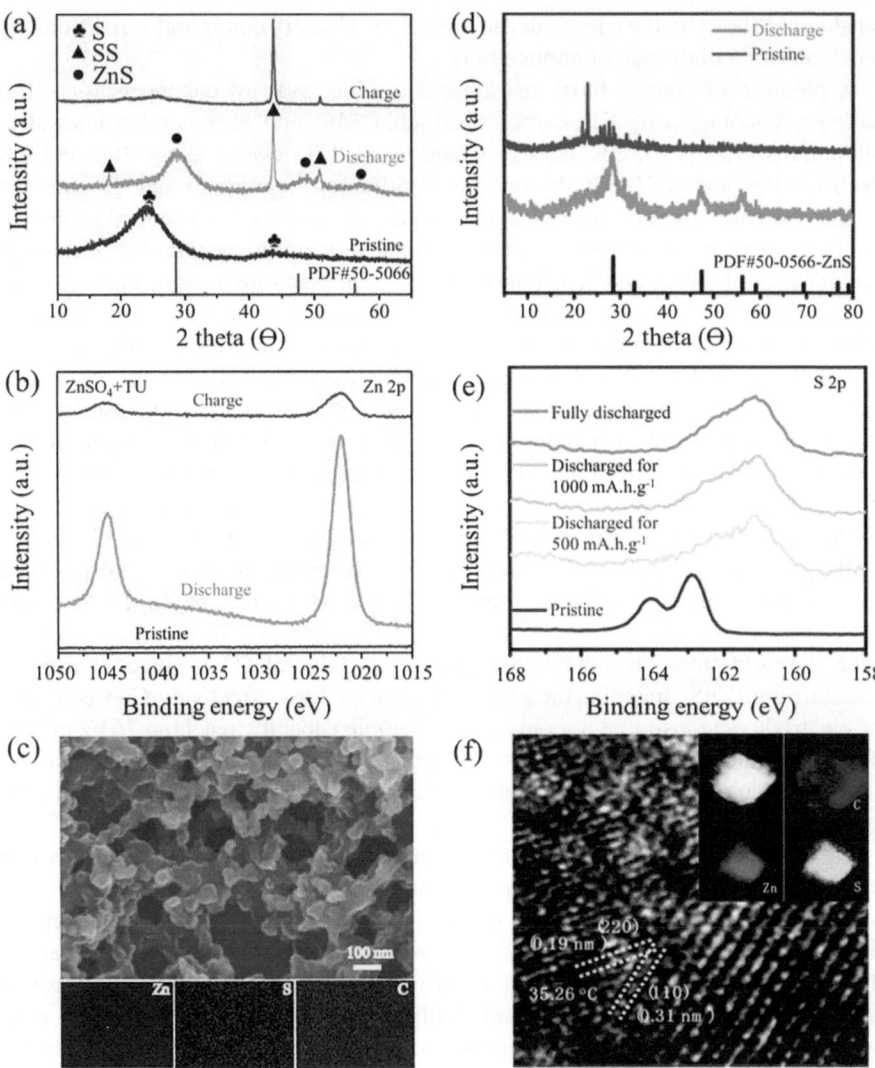

Fig. 3.20 Investigation into the charge and discharge mechanisms of S@KB cathode, Chang et al. [7]. **a** XRD analysis of the S@KB electrode to observe changes across various charge and discharge states. **b** Spectral analysis of Zn 2p for the S@KB electrode was performed during various charge and discharge states. **c** SEM micrographs examining the completely discharged state of the S@KB electrode [7]. **d** XRD analysis was conducted on both the pristine and fully discharged sulfur cathodes. **e** XPS spectra captured for sulfur cathodes at various SOCs, while maintaining a discharge rate of 50 mA/g. **f** Image of the discharge sulfur cathode captured by HR-TEM. Reproduced with permission from [7,13]

outstanding results, showcasing a substantial capacity, reaching 1,668 mAh/g and a prominent discharge plateau at approximately 0.7 V. Additionally, ex situ analysis provided insights into direct transformation mechanisms from sulfur to ZnS, emphasizing the fundamental chemistry involved in the aqueous Zn–S system. It is worthwhile to mention that a notable feature of the sulfur cathode was its remarkable performance at substantial mass loading levels, reaching up to 8.3 mg, yet preserving a specific capacity of 1,375 mAh/g. Consequently, this equated to an areal capacity of 11.4 mAh/cm^2 and an areal energy of 7.7 mWh/cm^2.

Considering sulfur's widespread availability and its cost-effectiveness, coupled with the inherent safety aspects of aqueous electrolytes, this chemical approach holds significant promise for making affordable, high-energy, eco-friendly primary batteries. In summary, the comprehensive strategy, integrating a finely manufactured electrode structure with a thoughtfully selected electrolyte additive, resulted in significant enhancements in reaction speed and enduring stability. Consequently, this investigation established a novel milestone in maximizing cathode efficacy, underscoring the reliability of melt-diffusion as a methodology and unveiling TU as an exceptionally potent additive within the domain of Zn–S batteries. Within the rapidly evolving realm of Zn–S batteries, the focus of innovation has shifted distinctly toward the cathode material. An investigation conducted by Gross and Manthiram [23] delved into the application of cobalt sulfide (CoS) as an additive aimed at augmenting cathode performance. By employing a modified technique tailored for cobalt electrodeposition, stainless steel, as well as brass meshes were opted as substrates. Following throughout treatments involving cleaning, soaking, and acid-based deoxidization, the meshes were immersed in a deposition bath containing cobalt S-3 acetate and boric acid. This bath facilitated the electrodeposition process, resulting in the transformation of cobalt to CoS. It is noteworthy that the advanced setup of ZAPS batteries integrated these meshes enhanced with CoS. What makes this battery special is its innovative design, utilizing a mediator-ion SSE to segregate the zinc anode from the highly reactive polysulfide catholyte. To preserve charge equilibrium, mediator ions, notably Na$^+$ or Li$^+$, were opted for within the system, chosen from the alkali metal category. This yielded a high-performance battery achieving a reversible capacity, reaching 822 mAh/g following 50 cycles, accompanied by an almost perfect CE of close to 100%. Noteworthy is the fact that ZAPS batteries incorporating Na$^+$ as the mediator ion outperformed their counterparts containing Li$^+$ in terms of rate and power measurements. To gauge the efficacy of these advanced batteries, a series of CV tests were conducted employing a dedicated battery cycling instrument. The examinations unveiled promising directions for future investigation, including strategies to improve the conductivity of the SSE and fine-tune catalyst effectiveness. To sum up, the integration of CoS-enhanced mesh cathodes and the architecture of ZAPS batteries offer a promising pathway in the pursuit of energy storage solutions that are both high in energy density and cost-effective, while also maximizing efficiency.

Zhang et al. [43] engineered a novel atomic iron catalyst bi-directionally to amplify the effectiveness of sulfur cathodes within Zn–S batteries. Using a cellulose wiper cloth readily accessible in the market as a starting point, the researchers proceeded by applying a layer of polypyrrole hydrogel, followed by subsequent chemical treatment, and annealing at high temperatures. Figure 3.21a visually outlines the procedure, featuring SEM images displaying the iron loaded with sulfur. The research findings highlighted that the iron sites, structured at the atomic level and coordinated with Fe − N_4, served as catalytic agents capable of operating bi-directionally. These catalysts played a crucial role in significantly accelerating the reversible sulfur conversion throughout the battery cycling process. To fabricate the sulfur cathode, a segment of a finalized product was amalgamated with sublimed sulfur within a glass vial. Following this, carbon disulfide was introduced, and the blend was agitated at a moderate temperature until complete dissolution of the sulfur and subsequent evaporation of the solvent occurred. Subsequently, the blend was transferred into a Teflon-lined container and exposed to a heating procedure consisting of two distinct steps.

As depicted in Fig. 3.21b, c, the expansive synthetic approach adopted to fabricate the sulfur/Fe and nitrogen-doped carbon on carbon cloth cathode, denoted as S@FeNC/NC/CC, signifies a revolutionary leap forward within the field of Zn–S batteries. The inventive core–shell configuration of the composite materials, highlighted clearly through SEM imaging, greatly amplifies the electrochemical prowess of the Zn–S battery system. In Fig. 3.21d, TEM analysis unveils supplementary understanding regarding the cohesive merging of the active carbon layer with the carbon fiber base, revealing a thickness of around 500 nm. In Fig. 3.21e, the FeNC/NC sheath is depicted, highlighting its amorphous texture, and providing convincing proof of the absence of noticeable iron clusters on its surface. A comprehensive examination employing advanced high-resolution High-Angle Annular Dark Field-Scanning Transmission Electron Microscopy (HAADF-STEM) techniques was undertaken to explore the configuration of iron contained within the carbon shell. As depicted in Fig. 3.21f, a multitude of scattered points, delineated by circled areas, display noticeable image contrast attributed to the heightened scattering effects of the transition metal constituents. The heightened contrast observed serves as a clear validation of the intricate incorporation of ultrafine iron single atoms into the nitrogen-doped carbon matrix. Figure 3.21g showcased the application of Energy-Dispersive X-ray Spectroscopy (EDS), which was employed to validate the distribution of compositions. After acid-washing treatments, the mononuclear iron atoms remained undisturbed, indicating the successful elimination of acid-vulnerable iron nanoclusters, resulting in the preservation of solely stable, pristine iron atoms. It is believed that these atoms cooperate in synergy with nearby nitrogen atoms, leading to the formation of sturdy iron–nitrogen structures that are resistant to acid. In particular, the matrix's robust interaction with sulfur species was instrumental in achieving nearly total sulfur-to-ZnS conversion throughout the discharging process of the battery. Concurrently, throughout the recharging process, the dynamic iron sites effectively diminished the activation energy barrier for ZnS, enabling a highly reversible process wherein electrodeposited ZnS transformed back into sulfur. Consequently, the newly engineered independent sulfur cathode displayed outstanding performance parameters, encompassing a reduced potential hysteresis measuring 0.61 V, a substantial

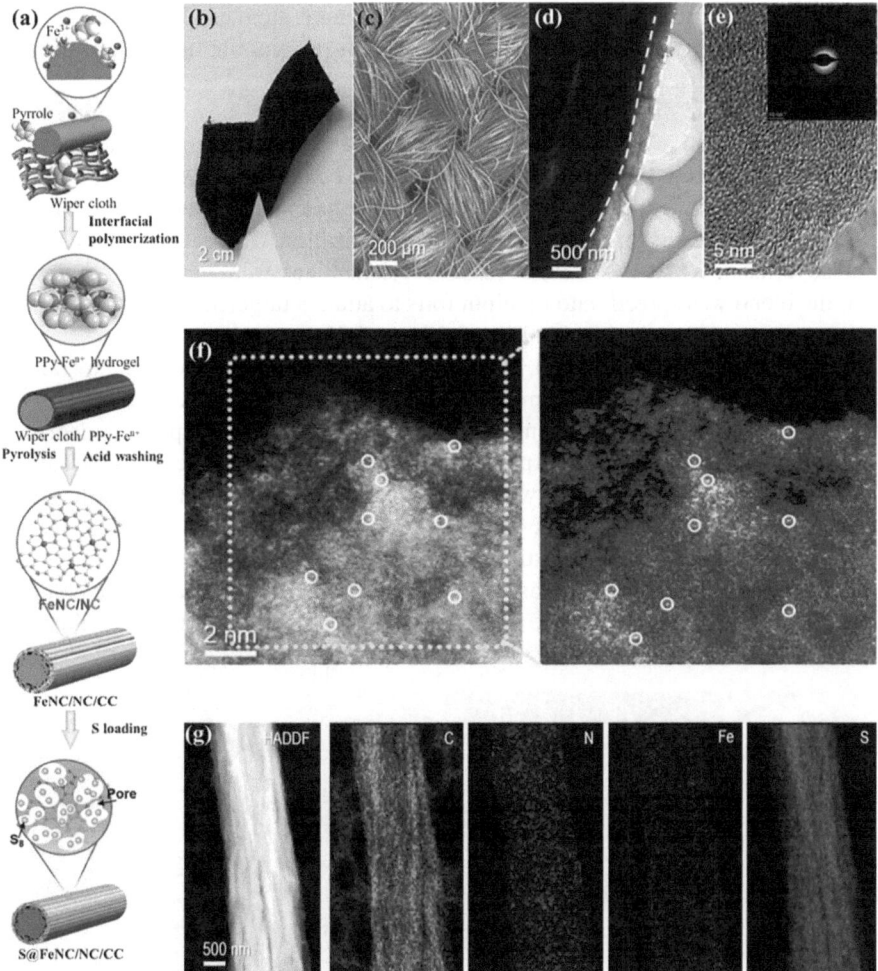

Fig. 3.21 Visual representation outlining the production process of an independent sulfur cathode alongside the distinctive structural features of the composite cathode. **a** Illustrative process map delineating the formation of S@FeNC/NC/CC through stages including polymer coating, carbonization, elimination of iron species, and subsequent loading of sulfur onto S@FeNC/NC/CC. **b** Digital image, **c** SEM micrograph, **d**, **e** TEM images, **f** atomic resolution HAADF-STEM depiction, and **g** associated elemental cartography of the sulfur-infused FeNC/NC/CC material. Reproduced with permission from [43]

discharge capacity reaching 1,143 mAh/g, and remarkable cycling durability maintaining a degradation rate of merely 0.141% per cycle over 300 cycles. Additionally, the investigation unveiled the deterioration process in these batteries to be the aggregation of dormant ZnS nanocrystals. Moreover, a successful demonstration of a ductile Zn–S battery with high performance was achieved. These discoveries not only shed light on the revolutionary capability of single-atom catalysts in controlling sulfur redox reactions but also indicate significant progress in advancing practical metal-sulfur batteries.

In the realm of Zn–S batteries, diverse research efforts have been dedicated to enhancing the active components and electrolyte chemistry of the cathode. The investigation conducted by Xu et al. [15] involved examining how CMK could be implemented on the sulfur cathode, with a focus on enhancing the mass loading of the active component on titanium foils for optimization. To construct the cathode, precise ratios of CMK-3@S, PVDF, and acetylene black were identified. Following that, the blend was spread onto titanium foils to attain a targeted mass loading of the active component, as depicted in Figs. 3.22 and 3.23.

The anode comprised zinc foil, while fiberglass functioned as the separator, alongside electrolytes specifically formulated with an iodine (I) additive. Battery performance underwent evaluation using GCD and EIS techniques, employing dedicated testing equipment and predefined parameters. To enhance the efficiency of the battery, tailored electrolytes containing I were utilized for further optimization. Furthermore, in conjunction with these endeavors, a comprehensive evaluation was carried out utilizing EIS and corrosion investigations.

Figure 3.22a, b display SEM and TEM micrographs, demonstrating the homogeneous incorporation of sulfur into the porous CMK-3 matrix. Figure 3.22c illustrates

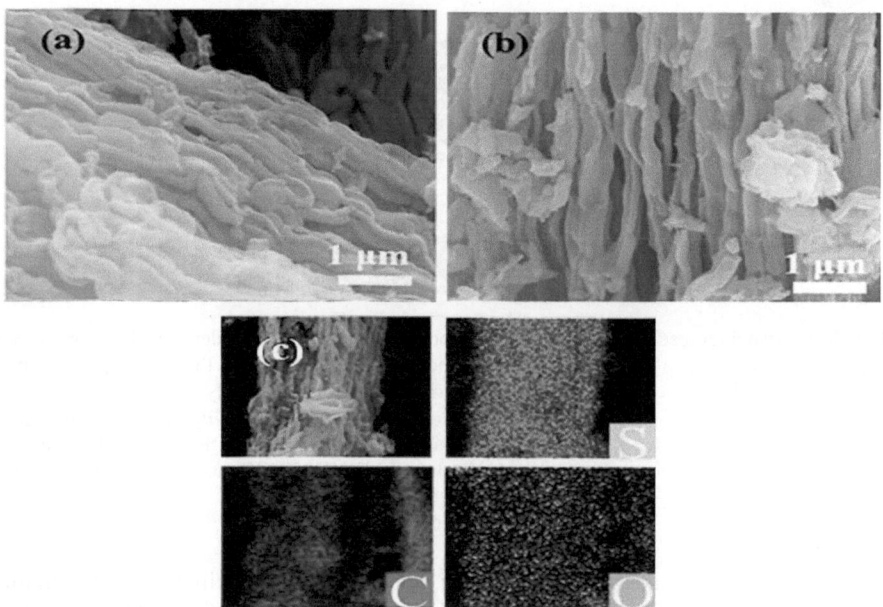

Fig. 3.22 Investigation into the composite structure of CMK-3@S. **a, b** SEM micrographs representing **a** CMK-3 and **b** CMK-3@S composite. **c** SEM visualization coupled with elemental mapping. Reproduced with permission from [15]

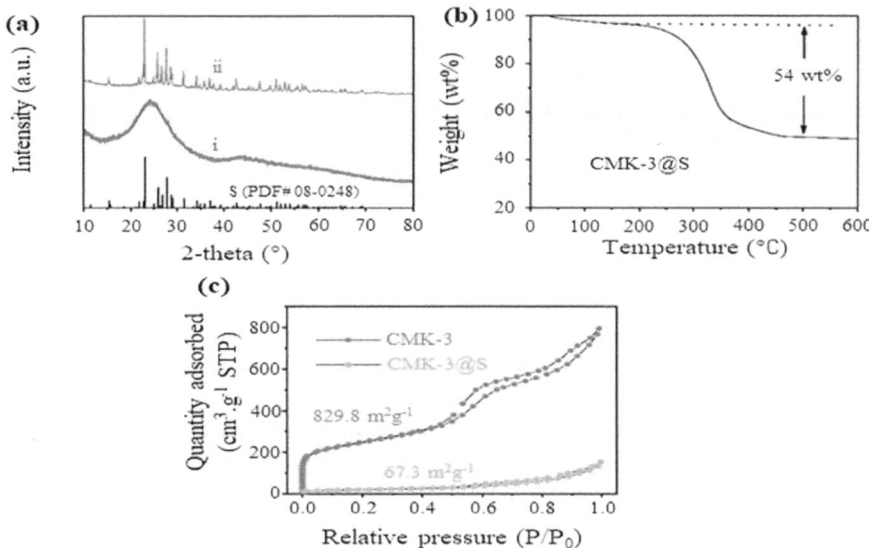

Fig. 3.23 Investigation into the composite structure of CMK-3@S. **a** XRD profile obtained from the CMK-3@S composites both prior to and after heat treatment at 155 °C (ii); **b** Thermal gravimetric analysis (TGA) profile displaying the sulfur content by weight in the CMK-3@S composite. **c** Profiles associated with nitrogen adsorption–desorption isotherms acquired from both CMK-3 and the CMK-3@S composite. Reproduced with permission from [15]

EDS mapping, offering supplementary evidence of the homogeneous encapsulation observed in the sulfur integration process within CMK-3. Further validation of this structure is provided by XRD patterns, as depicted in Fig. 3.23a. The diffraction peaks observed at 23, 25, and 27 degrees are conclusively associated with elemental sulfur, as confirmed by the PDF# 08-0248 standard. An intriguing observation is that the peaks representing bulk sulfur almost disappear after integration into CMK-3. These findings imply a dual phenomenon: sulfur spreading within the CMK-3 structure and then being contained within its inner pores, culminating in an amorphous state with lower sulfur crystalline properties. The weight percentage of sulfur content into the modified CMK-3, known as CMK-3@S, was found to be 54 wt.%, as illustrated in Fig. 3.23b. Additional affirmation of the effective encapsulation of sulfur is gleaned from the nitrogen adsorption–desorption isotherms, as depicted in Fig. 3.23c, showcasing a notable decline in specific surface area, plummeting from 829.8 m^2/g in the case of pristine CMK-3 to 67.3 m^2/g for CMK-3@S. The alternation in surface area, coupled with the concurrent adjustment in pore size distributions, provides compelling evidence of the effective amalgamation of sulfur into the CMK-3 framework. This suggests a promising prospect for a considerable boost in the specific capacity of Zn–S batteries. The electrolyte chemistry stands out as an area of special interest, exerting a pivotal influence on the functioning of Zn–S batteries. Through methodical examinations, it has been determined that $Zn(OTF)_2$ surpasses both $Zn(AC)_2$ and $ZnSO_4$ in terms of performance among the electrolytes studied. This superiority is mainly attributed to the presence of larger CF_2SO_3 anions, which significantly affect the solvation shell structure of Zn^{2+}. Of note is the remarkable attainment of reaching a reversible capacity of 788 mAh/g in concentrated electrolyte containing 3 M $Zn(OTF)_2$. This success can be credited to the heightened overpotential for zinc nucleation and the slowed-down pace of corrosion reactions. These impressive outcomes were obtained through the employment of S@CMK cathodes.

Collectively, the findings from these investigations not only push forward our knowledge of the ideal setups for cathodes but also emphasize the crucial impact of electrolyte chemistry on the functioning of batteries and their lifespan.

Liu et al. [31] proposed an innovative strategy that combines two-dimensional and three-dimensional Zincophilic Mesoporous Sieving techniques. Figure 3.24 illustrates some outcomes. The choice of solvent differed according to the dimensional version, leading to variations in the procedures. In both scenarios, the outcome was a light-yellow powder, achieved by effectively removing any remaining surfactant via high-temperature treatment in a tube furnace.

Figure 3.24 offers valuable insights into how changes in pore size influence the electrochemical performance of the cathode within Zn–S batteries. In terms of morphology, the GCD diagram of the two-dimensional zinc/sulfur (2DZS) distinctly reveals a plateau segment within the curve recorded during the discharging process, depicted in Fig. 3.24a. According to the measured data indicated in Fig. 3.24b, it is

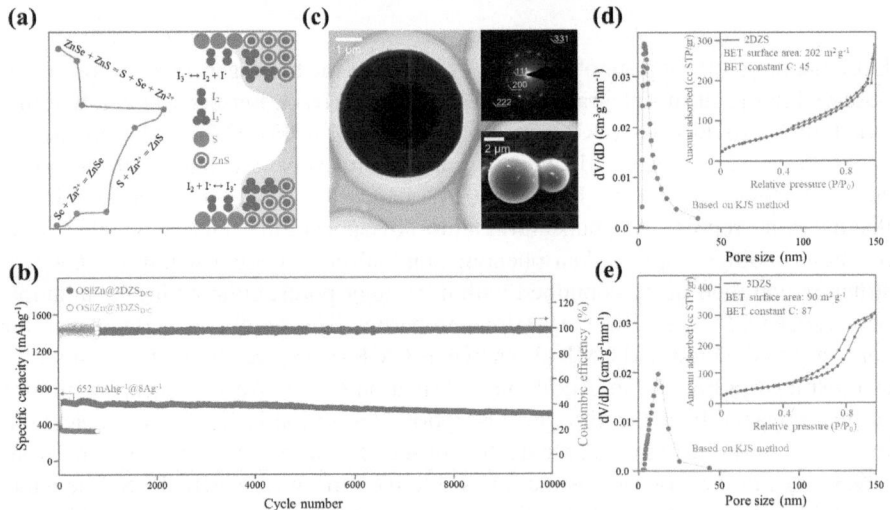

Fig. 3.24 **a** GCD profile representing the electrochemical behavior of the Zn–S battery with the inclusion of selenium (Se) and I_2 additives. **b** Evaluation of high-rate cycling performance conducted on the OS||Zn@2DZS and OS||Zn@3DZS batteries under 8 A/g discharge rate. **c** SEM and TEM-micrographs pertaining to 3DZS specimens. **d, e** Distribution of pore size within **d** 2DZS and **e** 3DZS specimens, coupled with measurements of the respective surface areas. Reproduced with permission from [31]

evident that a battery employing the optimized Zn@2DZS anode showcases extraordinary durability, surpassing 10,000 cycles. Concurrently, at a high discharge rate of 8 A/g, the battery represents a specific capacity measuring 652 mAh/g, with calculations made relative to the sulfur-based cathode. The pore size distribution, denoted as PSD, assessed via the Kruk-Jaroniec-Sayari (KJS) methodology, is utilized to quantify the mesoporous nature of these materials. By giving priority to the analysis of the adsorption isotherm, this methodology effectively removes any artifacts originating from tensile strength effects. The highest PSD value measured in the 2DZS specimens is a modest 4 nm, which is considerably lower compared to the 11 nm measured in the 3DZS specimens. This observation provides further evidence for the theory, which was substantiated by earlier atomic force microscopy (AFM) findings, indicating that 2DZS harbors accessible mesoporous that promote effective electrochemical interactions. Figure 3.24d, e provide detailed insights into this comparative investigation, illustrating a convincing relationship between pore size and the battery's electrochemical performance. This investigation holds significance as it indicates that manipulating pore size could potentially enhance the specific capacity across different categories of Zn–S batteries, an area that has received limited attention in previous research.

Summary: In the pursuit of enhancing sulfur cathodes in Zn–S batteries, researchers are exploring alternatives beyond conventional carbon additives. Both Chang and Luo utilized melt-diffusion methodologies with KB to produce electrodes, resulting in high-performing outcomes. Gross and Manthiram examined the utilization of CoS as an additive, employing an innovative electrodeposition process on stainless steel and brass meshes. Zhang et al. [43] revolutionized cathode efficiency with the introduction of a groundbreaking bi-directional atomic iron catalyst, while Xu et al. [15] delved into performance evaluation by utilizing CMK alongside custom I-infused electrolytes. Liu et al. [31] proposed a distinctive methodology incorporating two-dimensional and three-dimensional Zincophilic Mesoporous Sieving, resulting in producing favorable outcomes. Taken together, these investigations signify promising strides in the development of electrode materials and the overall enhancement of battery performance.

3.4 Fabrication, Advancement, and Implementation of Zinc-Derived Anodes

The reversible mechanism governing charge storage in zinc anodes within Zn–S batteries revolves around the deposition and stripping of zinc ions, a notion corroborated by multiple citations in the literature [48]. As the charging process progresses, zinc ions are reduced and subsequently adhere to the zinc anode's surface. Conversely, as the discharging process progresses, the zinc ions previously attached to the anode's surface are removed and converted back into soluble zinc ions in the aqueous medium. The presence of the electric field within the electrolyte prompts the migration of the soluble zinc ions toward the cathode [36, 49]. It is worth emphasizing that the process of depositing and stripping zinc ions occurs across multiple regions on the surface of the zinc anode, rather than being restricted to one area. The zinc ions tend to aggregate primarily at screw dislocations located on the surface of the anode, influenced by how these dislocations affect the distribution of the electric field. Consequently, this process culminates in the creation of zinc dendrites [50, 51]. The proliferation and growth of these zinc dendrites amplify a range of secondary reactions, such as corrosion and the HER, as they augment the zinc anode's surface area that is exposed. The Zn–S battery's anode grapples with hurdles like passivation, which arises from the deposition of ZnS, and corrosion, triggered by polysulfides. The onset of ZnS passivation initiates with the deposition of the discharge by-product, known as ZnS, onto the anode's surface. This process creates a hindering barrier that obstructs the transport of both ions and electrons. Consequently, this leads to a reduction in the efficiency and capacity of the battery [44]. The presence of polysulfides, a frequent occurrence in sulfur-based batteries, can instigate corrosion, ultimately resulting in the deterioration of anode material and consequently undermining the efficiency of the battery [15, 31]. This leads to diminished battery capacity, heightened impedance, and triggers an unfavorable cycle where the expanded interface fosters the creation of by-products and nucleation sites for more zinc dendrites. Ultimately, this cascade of events leads to the failure of the battery [52, 53]. Fundamentally, the performance of the zinc anode is profoundly influenced by its structural traits. Therefore, it makes sense to improve the performance of the anode by making alternations to the foundational structure of the zinc anode across all zinc-ion ESSs. In general, the fundamental guidelines for developing anode materials in diverse classifications of ZIBs involve a multitude of key considerations:

(a) Establishing a reliable support structure that can withstand the extended plating and stripping processes of zinc ions over time, without encountering structural failure or substantial changes in shape [54].

(b) Developing an extensive network of interconnected tunnel pathways and voids to enable the swift movement and retention of ions and anions [55].

(c) Providing a large surface area with active sites to increase the interaction between the anode and electrolyte, facilitating consistent zinc deposition [56].

(d) Exhibiting exceptional electrical conductivity and hydrophilic properties to promote the penetration of electrolytes and facilitate thorough cycles of plating and stripping [57].

(e) Displaying durability and chemical resilience in both acidic and alkaline electrolytes, while also having the ability to develop a strong SEI on the surface [58].

Despite the effectiveness of approaches such as benefiting from 3D anodes, utilizing zinc alloy anodes, and integrating additives into the anode to enhance the electrochemical features of Zn–S batteries. It is essential to mention that the development of Zn–S batteries is at an early stage, emphasizing the critical importance of ongoing research efforts. In the past, zinc foil has commonly been utilized as the preferred material for anodes. However, in specific scenarios, the presence of sulfate ions has expedited the corrosion process. In the modern era, there have been notable breakthroughs in the fields of materials science and nanotechnology, resulting in the introduction of innovative advancements. Such advancements encompass the development of pure zinc-based 3D anodes, as well as the improvement of zinc anodes through the incorporation of diverse functional energy materials like transition metal elements, metal oxide matrices, and structures based on carbon components.

The literature explores various methodologies aimed at tackling the difficulties related to non-uniform deposition and growth of zinc metal throughout cycling. These difficulties encompass the formation of the dendritic structure of zinc, internal short-circuiting, increased corrosion rate, the generation of inactive zinc components, and achieving less-than-optimal CE [59–62]. Included in these methodologies are alternations to the interfacial electric field [63–66], modification to the environment surrounding the zinc coordination [42, 67–73], and control over the processes involving zinc deposition [74–77] among various other methodologies explored. One particularly effective methodology involves the implementation of a protective film, such as a CNG membrane [78], Ag [79], ZnO [80], $CaCO_3$[81], and others [82–84], which serves to enhance the interface between the zinc anode and aqueous electrolyte, thereby promoting the stability of the zinc metal anode. The presence of this protective film effectively suppresses undesirable side effects, encompassing the corrosion of zinc metal, the generation of ZnO, and the occurrence of the HER [51].

An alternative highly effective technique for curtailing the growth of zinc dendrite structures is the application of electronically conductive 3D scaffolds. These scaffold structures contribute to the uniform distribution of electric fields, adaption to volume fluctuations, and decrease concentrated current density throughout the zinc plating and stripping procedures [77, 85, 86]. In this context, research endeavors have revealed the effectiveness of 3D MXene materials, renowned for their outstanding hydrophilic features and conductivity, which play a pivotal role in facilitating consistent zinc deposition and ensuring consistent cycling performance [87, 88].

Nevertheless, notwithstanding these progressions, overcoming the hurdles for the commercialization of rechargeable zinc-based energy storage technologies remains a formidable task. The expectation is that by amalgamation of protective layers with electronically conductive 3D scaffold structures and ionically conductive films, there will be a considerable boost in electrochemical efficiency. Remarkably, scant documentation exists within pieces of literature regarding this integrated methodology.

Considering this, a thorough examination has been initiated of recent progressions in the advancement of high-performance zinc anodes intended for Zn–S batteries. Most research endeavors primarily concentrate on refining cathode and electrolyte constituents to mitigate polarization effects, often disregarding the challenges surrounding the anode and its corresponding polarizations. Despite the scant endeavors dedicated to enhancing the anode, these initiatives highlight the pivotal importance of tackling difficulties pertaining to the anode to elevate the overall effectiveness of Zn–S batteries. Hence, there is only a handful of investigations regarding modified zinc anodes that are described in the sections below.

3.4.1 Zinc Foil Anode and Its Corresponding Anodic Hurdles

Guo et al. [18] undertook a study examining how cycling affects zinc foil in a Zn–S battery setup. They carried out experiments involving zinc stripping and plating in diverse electrolyte solutions to examine how hybrid electrolytes affect the suppression of zinc dendrite formation. The investigations encompassed the utilization of both Zn/Zn symmetric cells and Zn/Ti asymmetric cells. Upon completion of 100 cycles, the researchers utilized SEM for morphology analysis and XRD for composition assessment of the zinc electrodes.

During preliminary phases, the Zn/Zn symmetric cells displayed a consistent potential profile throughout zinc stripping and plating within a 2 M $Zn(CF_3SO_3)_2$ solution dissolved in deionized water, denoted as AE. Yet, this consistent performance waned following 220 h of cycling, possibly because of elevated resistance stemming from the growth of the dendritic zinc structure. Unexpectedly, Zn/Zn symmetric cells, when subjected to zinc stripping and plating in hybrid electrolytes consisting of 2 M $Zn(CF_3SO_3)_2$, supplemented with a ZnI_2 additive, and EG making up 50% of the total volume (HE-50), exhibited a prolonged lifespan, maintaining a consistent potential profile, surpassing 1,400 h. Conversely, the zinc stripping and plating of Zn/Ti cells in AE experienced a swift decline in CE during the first 100 cycles. This decline stemmed from the proliferation of zinc dendrites and accompanying secondary reactions involving water. The prolonged cycle durability and enhanced CE unequivocally demonstrated the role of the hybrid electrolyte in promoting the reversibility of the zinc stripping and plating, concurrently inhibiting zinc dendrite formation, effectively [18].

To provide additional evidence of the hybrid electrolyte's capability to suppress dendrite formation, researchers examined zinc electrodes extracted from Zn/Zn symmetric cells after subjecting them to 100 stripping and plating cycles within both AE and HE-50 electrolytes using post-mortem SEM and EDS analysis. This investigation is depicted in Fig. 3.25a–d. Contrasting the non-uniform, dendrite-ridden surface of the zinc electrode cycled in AE (as seen in Fig. 3.25a), the zinc electrode that underwent cycling in HE-50 displayed a noticeably smoother surface. According to Fig. 3.25b, a sleek surface was discernible on the zinc electrode immersed and cycled in HE-50 electrolyte. Furthermore, cross-sectional examinations depicted in Fig. 3.25d validated the existence of a compact SEI layer on the zinc electrode. Based on EDS examination, the compact SEI layer mainly comprised of elements such as Zn, C, and O, evenly dispersed across the surface of the zinc electrode (Fig. 3.26a). XRD analysis further confirmed the lack of dendritic formation in the zinc deposition process when using the HE-50 electrolyte, as depicted in Fig. 3.26b.

The XRD results closely mirrored those of the pristine zinc electrode, indicating uniformity in zinc deposition. It is worth mentioning that the SEI layer did not exhibit any recognizable XRD pattern, presumably owing to its amorphous composition. Conversely, the zinc electrode subjected to cycling in the AE electrolyte revealed identifiable peaks representative of basic zinc salts that had formed on its surface [18].

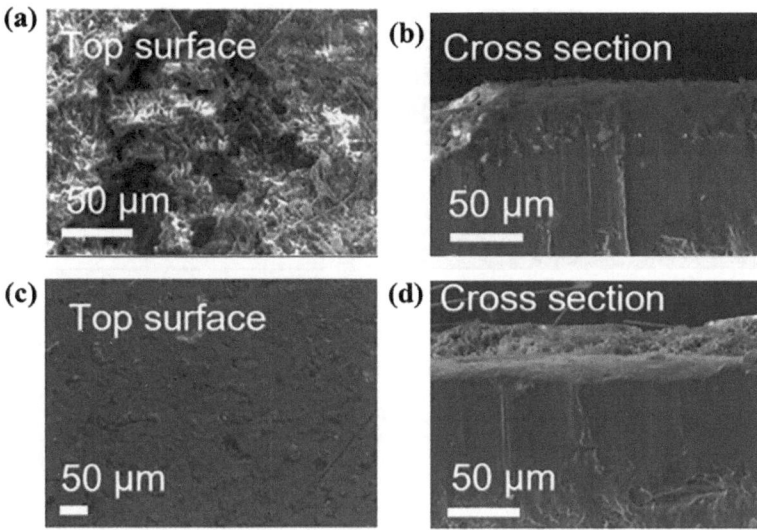

Fig. 3.25 a–d SEM micrographs portray the state of zinc electrodes after undergoing 100 cycles in both AE and HE-50 electrolytes. Reproduced with permission from [18]

According to Fig. 3.26c, EIS examinations provided further evidence supporting the constructive role of the SEI film in bolstering the durability of the zinc electrode. The impedance of the Zn/Zn symmetric cell within the AE electrolyte surged almost five times after subjecting it to 100 cycles of zinc stripping and plating, in contrast to the measurements obtained after the 10th cycle. On the other hand, within the HE-50 electrolyte, the charge-transfer resistance of the Zn/Zn symmetric cell consistently stayed at a low level, sustaining stability around 10 Ω. Consequently, the SEI film developed in HE-50 electrolyte functioned as a defensive shield, effectively curbing the formation of zinc dendritic structures and adverse side reactions within the aqueous electrolyte throughout the process of zinc stripping and plating [18].

According to results obtained from density functional theory calculations, denoted as DFT, it was discerned that when Zn^{2+} ions interact with EG and trifluoromethane sulfonate anions (OTF^-), there is a discernible decrease in the energy levels of the lowest unoccupied molecular orbitals (LUMO) pertaining to EG and OTF^-. These energy levels shift from their initial values of 0.74 eV and 0.67 eV to new values of $-$ 1.81 eV and -1.56 eV, respectively. The diminish in the LUMO energy levels within the assemblies' involving cations, anions, or solvents suggests an amplified eagerness to receive electrons from the anode, thus indicating a heightened inclination toward reduction. Therefore, employing DFT computations, the reduction potentials (versus. Zn^{2+}/Zn) for a variety of entities were anticipated, encompassing free solvent EG ($-$ 3.415 V), unbound anion OTF^- ($-$ 3.409 V), $Zn^{2+}-EG$ (0.27 V), and $Zn^{2+} - OTF^-$ (0.075 V). To sum up, the hybrid electrolyte has proven its effective capability to curb adverse and unwanted reactions involving sulfur at the cathode,

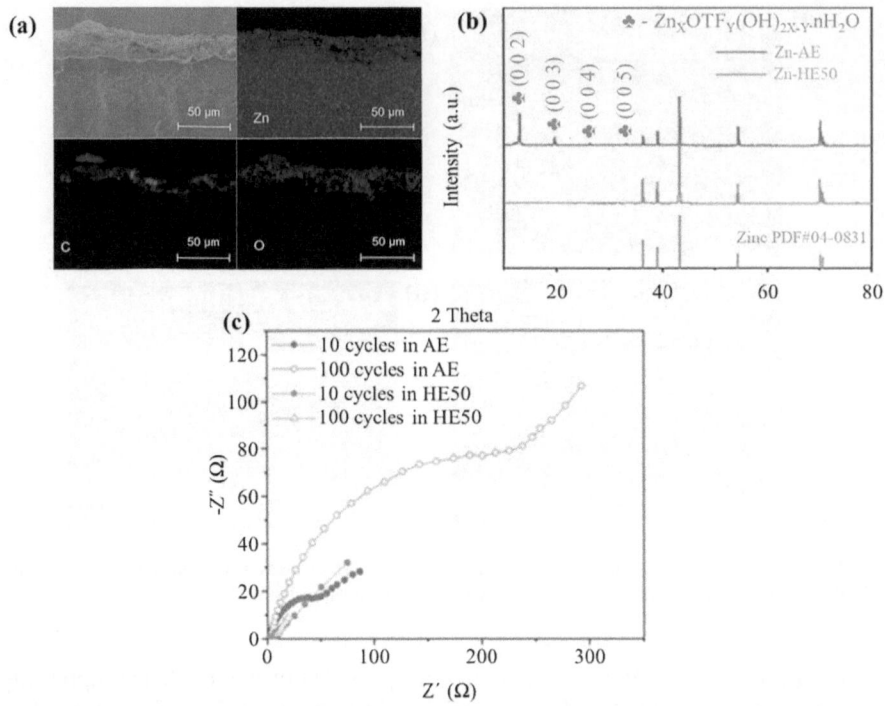

Fig. 3.26 a XRD spectra for the zinc electrode following 100 stripping/plating cycles within both AE and HE-50 electrolytes. **b** SEI film on the zinc electrode immersed within HE-50 electrolyte underwent elemental mapping employing EDS analysis. **c** EIS measurements conducted on Zn/Zn symmetric cells upon completing 10 and 100 cycles at a current density of 1 mA/cm^2 immersed within AE and HE-50 electrolytes. Reproduced with permission from [18]

as well as preventing the proliferation of zinc dendrites on the surface of the zinc metal anode. Moreover, functioning as a co-solvent within the hybrid electrolyte, EG possesses the capability to establish hydrogen bonds with unbound water molecules, thus proficiently alleviating adverse and undesirable side reactions linked to water molecules [18].

Xu et al. [15] devised a methodology employing $CF_3SO_3^-$ anions to control the corrosion reactions occurring on zinc anodes. Their investigation unveiled that within electrolytes of 1 M, 2 M, and 3 M Zn(OTF)$_2$ concentrations, the zinc nucleation overpotentials were measured at 70.7 mV, 100.4 mV, and 124.0 mV, respectively, as delineated in Fig. 3.27a. Therefore, they inferred that there is a direct correlation between the increase in electrolyte concentration in Zn–S batteries and the elevation in the zinc nucleation overpotential. Additionally, the augmentation in the concentration of electrolyte yielded a shift toward a more positive corrosion potential (E_{corr}) and a decline in corrosion current (I_{corr}), as indicated by the data presented in Fig. 3.27b, suggesting a reduced propensity for the corrosion of zinc electrode.

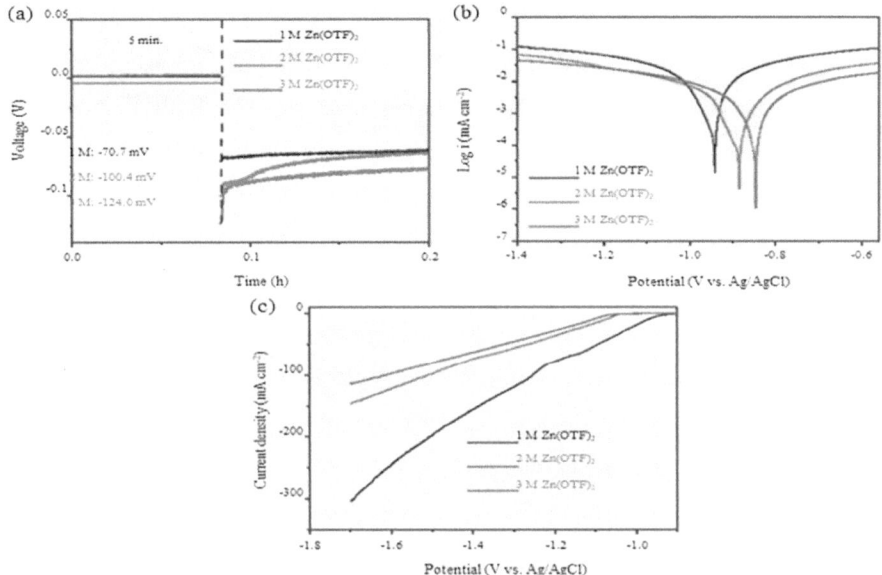

Fig. 3.27 **a** Overpotential necessary to trigger the onset of zinc nucleation utilizing various concentrations of Zn(OTF)$_2$. **b** Linear polarization curves representing electrochemical behavior of zinc anode in different concentrations of Zn(OTF)$_2$. **c** Linear polarization profiles portraying the response of the zinc anode throughout the HER process. Reproduced with permission from [15]

Furthermore, based on the HER assessment outcomes depicted in Fig. 3.27c, a reduced linear potential of -1.08 V was observed within 3 M Zn(OTF)$_2$ electrolyte, highlighting an expanded electrochemical window. This observation concurred with observations from SEM images, as illustrated in Fig. 3.28a–c, where zinc anodes immersed in dilute electrolytes displayed considerable corrosion and an elevated formation of zinc dendritic structures. On the flip side, a significant alternation was observed in concentrated electrolytes (3 M), where the surfaces of zinc anodes exhibited a dense and sleek appearance, free from any discernible protrusions. In conclusion, it was established that the consistent movement of Zn^{2+} ions in concentrated electrolytes facilitated uniform processes of zinc stripping and deposition, leading to the formation of a sleek and densely adherent surface of zinc metal.

In a distinct study, Yang et al. [44] developed a custom-formulated electrolyte, termed as a "cocktail-optimized" solution. This electrolyte formulation incorporated G$_4$ and H$_2$O as co-solvents alongside the inclusion of the additive I$_2$. This novel amalgamation of G$_4$ and I$_2$ displayed a synergy, effectively triggering the activation of the polar I$_3$/I catalyst pair while concurrently supporting the stability of the zinc anode through the facilitation of an organic–inorganic interphase formation throughout the cycling process. The lack of stability at the interface between the zinc anode and aqueous electrolyte on the anode side might instigate adverse side effects, encompassing the HER and corrosion, which could consequently lead to the formation of zinc dendritic structures. Furthermore, the inclusion of a sulfur cathode

Fig. 3.28 a–c SEM micrographs displaying the status of the zinc anode in Zn//Zn cells employing varying electrolyte concentrations, encompassing 1 M, 2 M, and 3 M, respectively. (The embedded micrographs present SEM views of the zinc electrode before undergoing the cycling procedure). Reproduced with permission from [15]

with substantial capacity could intensify the process of deposition and removal of Zn^{2+} ions in a single cycle, thereby compounding the hurdles encountered by the zinc anode even further.

To evaluate the longevity of Zn‖Zn symmetric cells through cycling process, researchers utilized 2 M $Zn(OTF)_2$ electrolytes, which comprised $Zn(OTF)_2$/water (Z/W), $Zn(OTF)_2/I_2$/water (Z/I/W), $Zn(OTF)_2/G_4$/water (Z/G/W), and $Zn(OTF)_2/$ G_4/I_2/water (Z/G/I/W). The experimentations involved testing these electrolytes at current densities of both 5mAh/cm^2 and 5 mA/cm^2, with the resulting data displayed in Fig. 3.29a. It is worth noting that the Z/I/W electrolyte-operated cell encountered a short-circuit occurrence after 12 h, a duration even briefer than that of the Z/W electrolyte cell, which endured for 18 h. This observation suggested that the inclusion of the I_2 catalyst expedited the corrosion process of the zinc metal and consequently hastened the degradation of the zinc anode. On the other hand, contrasting outcomes emerged with the inclusion of electrolytes containing G_4. Within the electrolyte containing Z/G/W, the longevity of the Zn‖Zn symmetric cell extended to 28 days. This prolongation can be predominantly ascribed to G_4's capacity to diminish the reactivity of water. Significantly, the incorporation of the I_2 additive into the G_4-infused electrolyte substantially expanded the cycling longevity of the Zn‖Zn symmetric cell to 75 days. These findings highlight the effectiveness of G_4 in efficiently protecting the zinc anode against the corrosive impacts of the I_2 additive. This deduction is supported by corrosion assessments conducted on zinc foils immersed in electrolytes containing Z/I/W and Z/G/I/W during periods of inactivity [44].

To shed light on this phenomenon, researchers conducted an extensive investigation into the electrochemical attributes of diverse electrolytes. As evidenced in Fig. 3.29b, electrolytes containing both Z/G/I/W and Z/G/W displayed notably elevated HER overpotentials in contrast to Z/I/W, highlighting the significant function of G_4 in suppressing HER. Moreover, while the surface of zinc anode subjected to cycling process in the electrolyte containing Z/I/W exhibited the signs of gradual roughening over time, the surfaces of those cycled in G_4-infused electrolytes (Z/G/W and Z/G/I/W) retained a comparatively smooth surface, as depicted in Fig. 3.29c. This implies that the inclusion of the G_4 co-solvent contributes to the inhibition of the formation and growth of dendritic structures, a conclusion substantiated by in situ optical microscopic examinations depicted in Fig. 3.29d. Furthermore, the zinc deposition patterns in diverse electrolytes were further examined employing chronoamperometry technique, denoted as CA, at a consistent overpotential of − 150 mV for a duration of 5 min, as demonstrated in Fig. 3.29e. Throughout the CA in the electrolyte containing Z/I/W, there was a continual increase in the current densities, implying the distribution of Zn^{2+} ions via the two-dimensional diffusion across the zinc surface. This, in turn, leading to an augmented surface area and thereby fostering the development of dendritic structures. Conversely, within the electrolyte containing Z/G/I/W, the current reached a gradual stabilization within a span of 25 s.

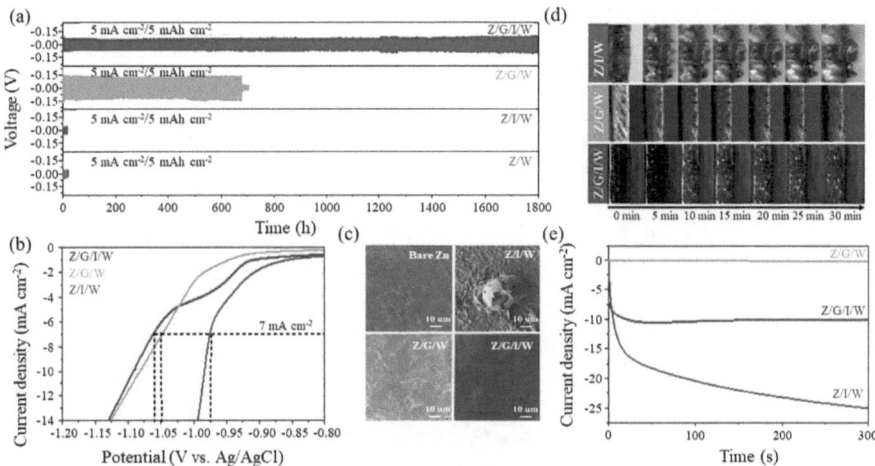

Fig. 3.29 Analyzing how the electrochemical behavior of zinc anode is affected by the improved electrolyte: **a** Evaluating the cycling efficacy within Zn‖Zn symmetric setups. **b** Linear sweep voltammetry (LSV) graphs depicting electrochemical behaviors employing multiple electrolytes under a potential scan rate of 10 mV/s. **c** SEM micrographs illustrating a comparative analysis between pristine zinc anode and following the cycling process. **d** Optical microscopic examinations of the deposition process of Zn^{2+} ions onto the zinc anode electrode. **e** CA plots depicting the zinc deposition patterns measured in diverse electrolytes containing Z/G/I/W, Z/G/W, and Z/I/W. Reproduced with permission from [44]

This suggests that the collaborative effect of G_4 and I_2 facilitated the emergence of extra zinc nucleation sites. This, consequently, promoted zinc growth by hindering the two-dimensional diffusion of Zn^{2+} ions and augmenting their diffusion kinetics. Of importance is the substantial disparity in the recorded current density observed in the Z/G/W electrolyte in comparison to both the Z/G/I/W and Z/I/W electrolytes.

This discrepancy could possibly be attributed to diminished ionic conductivity, leading to a less effective zinc deposition process within the electrolyte containing Z/G/W. This investigation unveiled that the collaboration between G_4 and I_2 promoted the development of an organic–inorganic hybrid SEI on the zinc anode. This, in turn, contributed to the stabilization of the zinc anode by mitigating undesirable and adverse interfacial side reactions [44].

Amiri et al. [33] conclusively exhibited that the corrosion could be expedited due to the development of SO_4^{2-} ions, coupled with the initiation of the OER. Efficient regulation of this phenomenon can be achieved by diminishing the ESPW. Figure 3.30a provides insight into the cycling attributes of a structural Zn–S battery functioning at a current density of 1 A/g, across an ESPW spectrum extending from 0.1 to −1.6 V.

At the outset, the crafted Zn–S battery showcased a discharge capacity of 511 mAh/g when subjected to 1 A/g current density, which progressively declined to 203 mAh/g after completing 300 cycles. Following cycles had a negligible effect on its

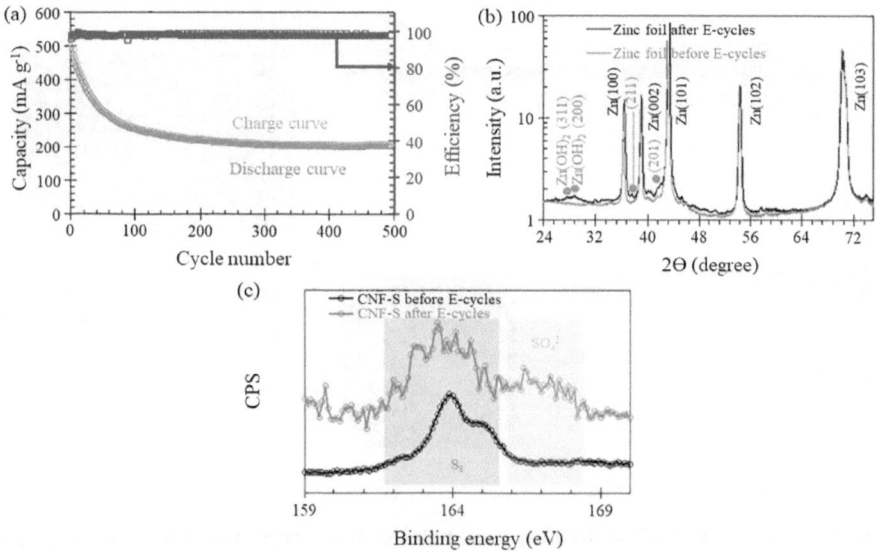

Fig. 3.30 **a** Charging and discharging behaviors and the corresponding CEs measured for the Zn–S battery operating at a current density of 1 A/g, covering the potential span from 0.1 to − 1.6 V. **b** XRD analysis conducted on the zinc foil prior and following completion of 500 E-cycles. **c** XPS analysis depicting the core level of S 2p in the CNF-S cathode both preceding and after 500 E-cycles. Reproduced with permission from [33]

discharge capacity and stayed consistent during the ensuing 200 cycles. Throughout the last 200 cycles, the Zn–S battery experienced a capacity degradation of around 3%, which was considerably less than the capacity degradation incurred during the initial 300 cycles. The significant degradation in the initial discharge capacity could be ascribed to a multitude of parameters, encompassing sulfur side reactions, complicated irreversible reactions associated with oxygen and hydrogen evolution, the emission of H_2S gas, as well as disproportionation and dissociation reactions occurring within the Zn–S battery. This investigation identified that the principal parameters contributing to the capacity degradation at higher potentials were the generation of SO_4^{2-} ions and the commencement of the OER. The XRD results depicted in Fig. 3.30b outline the zinc anode's structural alternations pre and post 500 cycles. According to Fig. 3.30c, upon the completion of 500 electrochemical cycles, the XRD results attained from the zinc foil revealed the appearance of multitude minor peaks at the discharge process, along with a sulfate peak appeared at S 2p. These observed peaks exhibited a strong correlation with the distinctive peaks associated with $Zn(OH)_2$. Additionally, the optical microscopy findings, as depicted in Fig. 3.31a, b, along with the observations made using confocal fluorescence microscopy, as illustrated in Fig. 3.31a, c, unambiguously indicate a considerable enhancement in the roughness of the zinc anode foil after 500 cycles. The confocal fluorescence microscopy observations reveal minuscule protrusions composed of small layers crafted form zinc; a detail corroborated by the SEM micrographs shown in Fig. 3.31c. The formation of these layers can be attributed to the deposition and subsequent stripping of Zn^{2+} ions. Furthermore, indisputable evidence for the partial oxidation of zinc surface elements after competing 500 electrochemical cycles was unequivocally highlighted through elemental mapping conducted on the zinc electrode's surface, where the coexistence of zinc and oxygen was clearly observed. In this scenario, the generation of SO_4^{2-} ions emerges as the predominant mechanism driving sulfur degradation and subsequent decline in the discharge capacity [33].

Various strategies can be implemented to alleviate zinc anode corrosion, encompassing the employment of diverse redox mediators. Liu et al. [3], developed a Zn–S setup featuring TUI additive as a redox mediator. TUI, containing I, functions as a proficient redox mediator, resulting in a decline in the charge overpotential, a trait previously observed in $Li–O_2$, Li–S, and Na–S batteries [89–91]. TU is also involved in modifying the crystallographic attributes and structure of the zinc surface, assisting in mitigation the emergence of dendritic structures and corrosion [92, 93]. Moreover, the formamidine disulfide cation exhibits the capacity to interact with I_3 ions, leading to an excess mitigation in the zinc anode corrosion.

Zhang et al. [32] proposed a methodology aimed at maximizing the utilization of Zn–S redox chemistry by managing the complicated relationship between kinetic reactions and the cathode composition. Throughout their inquiry, they formulated a high-energy Zn–S setup, where they developed a cathode exhibiting enhanced discharge capacity via in situ interfacial polymerization of $Fe(CN)_6^{4-}$-doped polyaniline within sulfur nanoparticles. In contradistinction to sulfur, the redox mediators $Fe^{II/III}(CN)_6^{4-/3-}$ showcased considerably accelerated cation incorporation and removal rates. The elevated cathodic potential ($Fe^{II}(CN)_6^{4-}/Fe^{III}(CN)_6^{3-} \sim 0.8$ V vs.

Fig. 3.31 Visual representation illustrating **a** the pristine zinc foil and **b** the zinc foil after undergoing 500 E-cycles. **c** SEM micrograph depicting the zinc foil following 500 E-cycles. The images illustrate three-dimensional visualizations acquired through confocal fluorescence microscopy for **a** the pristine zinc foil and **c** the zinc foil after subjecting to 500 E-cycles. Reproduced with permission from [33]

$S/S^2 \sim 0.4$ V) led to the spontaneous catalysis of sulfur's full reduction throughout the battery discharging process ($S_8 + Zn_2Fe^{II}(CN)_6 \leftrightarrow ZnS + Zn_{1.5}Fe^{III}(CN)_6$, $\Delta G = -24.7$ kJ/mol).

The inclusion of accessible iron redox species diminished the energy obstruction for ZnS activation throughout the reverse charging process, while the effortless migration of Zn^{2+} ions enabled a highly reversible transformation between sulfur and ZnS. Through a sequence of ex-situ examinations, the degradation mechanisms within Zn–S setups were elucidated, emphasizing that the clustering of inert ZnS nanocrystals was the principal culprit, as opposed to the depletion of the zinc anode. In addition to identifying the clustering of inert ZnS nanocrystals on the surface of the zinc anode as the principal culprit, it was acknowledged as the pivotal parameter hindering the reversibility of the active constituents of sulfur.

In tackling safety issues arising from the formation and growth of dendritic structures, as well as the occurrence of adverse and unfavorable side reactions, which impede the effective use of zinc anodes, An et al. [59] introduced a novel methodology centered around developing a protective heterogeneous film. Comprising two essential constituents, the protective film encompasses sulfur-doped three-dimensional (3D) MXene for electron conduction and ZnS for ion conduction. These constituents are applied to the zinc anode, leading to the creation of the composite structure referred to S/MX@ZnS@Zn.

In this design, the inclusion of sulfur-doped 3D MXene assumes a pivotal function in ensuring uniform electric field distribution, alleviating localized current density, and adapting volumetric fluctuations effectively [77, 86]. Concurrently, ZnS plays a role in improving the distribution of Zn^{2+} ions, which in turn promotes the movement and deposition of Zn^{2+} ions, thereby effectively inhibiting the creation of dendritic structures in the long run [64, 73, 94]. Consequently, through the implementation of this methodology, a dendrite-free anode structure of S/MX@ZnS@Zn is effectively synthesized. Investigation into the morphological transformation of zinc deposition was conducted on both unmodified zinc and the S/MX@ZnS@Zn-350 anodes, as visualized in Fig. 3.32. At the outset, the unmodified zinc anode displayed the emergence of minuscule zinc flakes, indicative of its zinc plating capacity, as depicted in Fig. 3.32b, b1. The emergence of these initial dendritic structures triggers the localized accumulation of electric charge, consequently promoting the growth of dendritic structures [87]. As zinc deposition progresses, at a rate of 2 mAh/cm^2 for instance, the ongoing proliferation of zinc dendritic structures persists, leading to the densification as observed in Fig. 3.32c, c1. As evidenced in Fig. 3.32d, d1, as the plating capacity elevated to 5 mAh/cm^2, zinc dendritic structures manifested in enlarged and increasingly porous configurations. Upon reaching a deposition capacity of 10 mAh/cm^2, zinc dendritic structures experienced heightened expansion, resulting in the formation of significant particles consisting of porous zinc dendritic structures, as depicted in Fig. 3.32e, e1. When the zinc plating capacity reached 20 mAh/cm^2, the proliferation of zinc dendritic structures exhibited irregular patterns, suggesting potential safety implications [64, 84, 95].

Additionally, following deposition, the unmodified zinc anode underwent a transformation into a rugged and permeable configuration, intensifying detrimental and unfavorable side reactions occurring at the interface between the zinc anode and electrolyte. These reactions encompass passivation, corrosion, and the emergence of by-products [96, 97]. Hence, the unmodified zinc anode exhibited subpar electrochemical efficacy when immersed within aqueous electrolytes. Employing this methodology has emerged as one of the most efficient approaches for the synthesis of a cathode customized for Zn–S batteries, adeptly tackling the hurdles associated with dendritic structures while bolstering the overall performance and safety aspects.

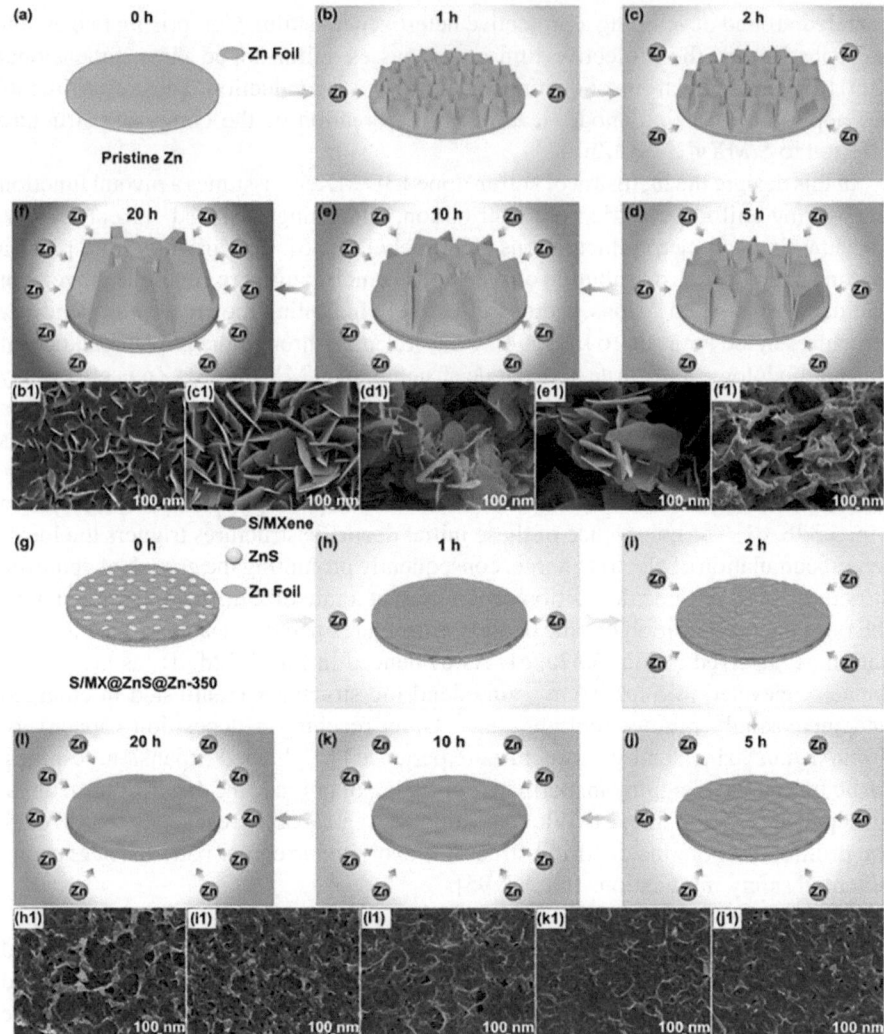

Fig. 3.32 Visual representation depicting the deposition of zinc ions on the surface of **a–f** unmodified zinc metal anode and **g–l** composite structural anode consisting of S/MX@ZnS@Zn-350 under a current density of 1 mA/cm^2 across varying deposition durations, featuring **a, g** no time lapse, **b, h** 1-h interval, **c, i** 2-h interval, **d, l** 5-h interval, **e, k** 10-h interval, and **f, j** 20-h interval. Investigation into the morphological transformation of **b1–f1** pristine zinc surfaces and **h1–j1** composite structural anode consisting of S/MX@ZnS@Zn-350 exposed to zinc deposition process under the current density of 1 mA/cm^2 across diverse deposition durations of **b1, h1** 1-h interval, **c1, i1** 2-h interval, **d1, l1** 5-h interval, **e1, k1** 10-h interval, and **f1, j1** 20-h interval. Reproduced with permission from [59]

3.4.2 Modified Zinc Anode and Its Corresponding Anodic Hurdles

While most research efforts prioritize optimizing the chain reaction occurring on the cathode's surface to mitigate polarization [98], the significance of anodic polarization is mostly undervalued. Consider, for example, the case was examining an 8 A/g rate is applied to the sulfur-based cathode; under these conditions, the associated current density at the zinc anode typically spans between 20 and 25 mA/cm^2. Therefore, a considerable polarization effect is evident throughout the entire Zn–S battery system, experiencing a voltage drop of 0.5 V–0.6 V. It is worth mentioning that even a slight polarization of 0.1 V has the potential to cause an energy density of around 150 Wh/kg_s [99]. Furthermore, the occurrence of reactions with competitive kinetics, encompassing I_2 corrosion, HER, and the proliferation of dendritic structures can heighten overpotential, ultimately culminating in the malfunction of the battery [14, 98, 100, 101]. Therefore, recent investigations have emphasized the critical need to engineer a zinc anode that minimizes polarization and mitigates susceptibility to competitive reactions, thereby paving the way for the practical implementation of Zn–S batteries [31, 47].

Li et al. [31] applied a methodology known as acid-assisted confined self-assembly, denoted as ACSA, to synthesize a two-dimensional (2D) mesoporous zincophilic sieve, termed 2DZS, which served as the kinetic interface. Following the ACSA procedure, the interface of 2DZS manifested a unique architecture of nanosheets extending into two dimensions, featuring plentiful zincophilic locations, possessing hydrophobic traits, and hosting minute mesopores. As a result, the 2DZS interface serves a dual purpose, addressing both nucleation and plateau overpotential: firstly, by accelerating the diffusion kinetics of Zn^{2+} ions along accessible zincophilic pathways, and secondly, by hindering competitive processes encompassing HER and the proliferation of dendritic structure due to the pronounced solvation–sheath sieving mechanism. Micrographs acquired from TEM and SEM were employed to visually assess the morphology of the synthesized 2DZS specimens, as illustrated in Fig. 3.33a. As a result, considerable improvement in electrochemical efficacy was achieved. To evaluate the polarization behavior more accurately, GCD curves presented in Fig. 3.33b validated that the overall polarization behavior of the synthesized Zn–S battery is merely 0.42 V at 1 A/g, notably less than most contemporary benchmarks. This suggests superior electrochemical attributes, as demonstrated in Fig. 3.33c. Hence, the decline in anodic polarization was evident, with a decrease to 48 mV at 20 mA/cm^2, contributing to the overall battery polarization that accounts for only 42% of that observed in a pristine Zn–S system. This led to the attainment of an exceptional energy density, measured at 866 Wh/kg at 1 A/g, along with an impressive durability expansion of 10,000 cycles even under the demanding conditions of a high rate of 8 A/g.

The emergence of dendritic structure and the initiation of HER on the surface of zinc anode present formidable hurdles to maintaining CE and ensuring long-term cycling stability of Zn–S battery systems. Addressing concerns associated with the

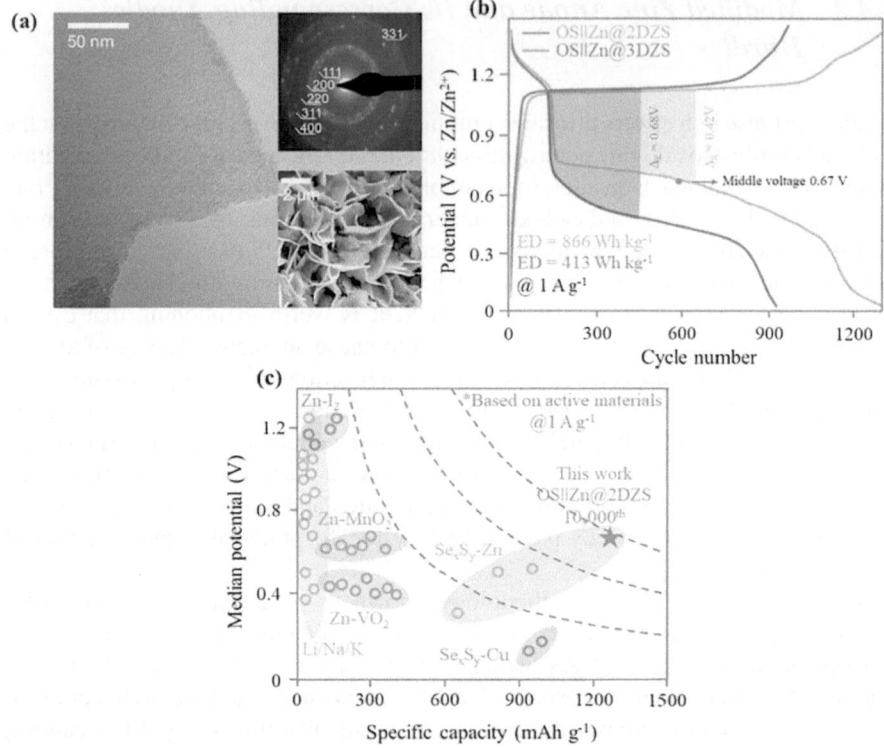

Fig. 3.33 **a** SEM, TEM, and selected area electron diffraction (SEAD) analyses conducted to investigate the structural characteristics of the 2DZS anode. **b** GCD curves employed to assess the polarization behaviors of the Zn–S battery featuring the 2DZS anode operating under a current density of 1 A/g. **c** The assessment of Ragone curve for various aqueous battery systems. Reproduced with permission from [31]

corrosion and emergence of dendritic structures on the anode's surface, Li et al. [47] innovatively engineered a resilient powder zinc/indium, denoted as pZn/In, anode that boasts both mechanical and chemical durability, specially tailored for Zn–S battery applications. This inventive architecture of anode integrated a coating of indium (In) onto the highly reactive pZn, fulfilling dual functions as a catalyst for nucleating and deterrent against corrosion, as demonstrated in Fig. 3.34a. Consequently, it effectively resolved concerns regarding both corrosion and dendritic structures, yielding superior utilization efficiency and prolonged cycling durability. Following the integration of the cathode based on sulfur with a negative-to-positive (N/P) capacity ratio nearing 2, the resultant full cell showcased an impressive initial specific capacity of approximately 803 mAh/g, as depicted in Fig. 3.34b. Throughout more than 300 cycles under a 2 C rate, it exhibited consistent performance with negligible capacity decay, averaging around 0.17% per cycle. Table 3.1 provides a concise overview of studies exploring various methodologies for enhancing Zn–S batteries, particularly in the realm of engineering novel anode materials.

Concluding our analysis and delving into the realm of anode materials designed specifically for Zn–S batteries, we can furnish a detailed and comprehensive exposition. Scientists have predominantly employed zinc foil as the primary material for the anode with Zn–S battery setups, yet there exists a potential for optimization through the investigation of alternative zinc anodes to improve the electrochemical performance. In the quest for optimal anode materials, it is imperative to satisfy numerous fundamental prerequisites:

(a) Ensure a durable architecture that can withstand prolonged cycles of zinc-ion plating and stripping without experiencing structural degradation or substantial changes in the morphological structure.
(b) Furnish a plethora of interstitial voids interconnected via tunneling networks to accelerate the conveyance and accumulation of ions and anions.
(c) Provide an expansive surface area featuring abundant active sites to promote effective interaction with the electrolyte, thereby ensuring even deposition of zinc ion.
(d) Ensure superior conductivity and hydrophilicity to promote electrolyte penetration efficiency and facilitate smooth plating and stripping processes.
(e) Maintain structural integrity and chemical stability when immersed in either acidic or alkaline electrolytes, while simultaneously having the capability to establish a resilient SEI layer on its surface. The emergence of SO_4 ions can intensify zinc corrosion within both aqueous and gel electrolytes, while the

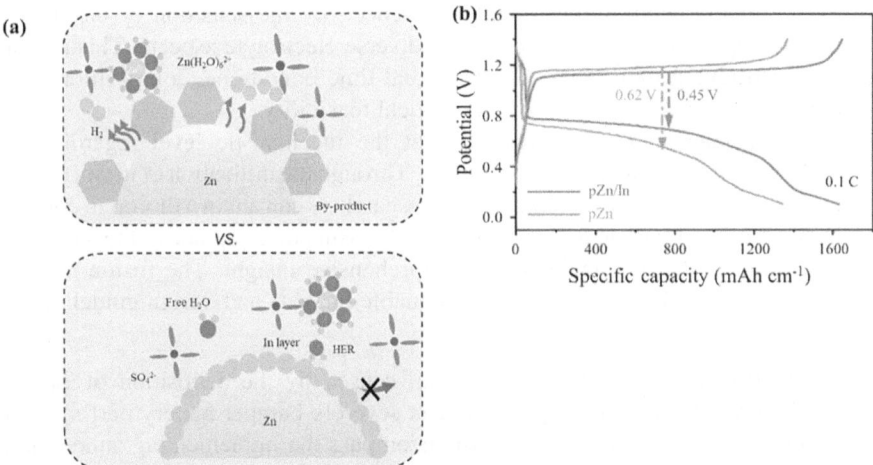

Fig. 3.34 a Schematic illustration exhibiting the corrosion mechanisms occurring within a Zn–S battery featuring a pZn/In anode. **b** Graph depicting the GCD profiles of a Zn–S battery employing a pZn/In anode at the rate of 0.1 C. Panels (**a, b**) were. Reproduced with permission from [47]

introduction of redox mediators presents a promising opportunity for enhancing surface protection.

In the realm of zinc-ion ESSs, anodes exhibit triple classification: (1) Three-dimensional (3D) variants distinguished by their considerably extensive surface area. (2) Anodes synthesized from zinc alloys. (3) Anodes enriched through the inclusion of supplementary additives. Although strides have been taken with zinc foil, there are lingering uncertainties regarding the development of a dependable zinc anode suitable for Zn–S batteries. Moving forward, it is imperative for research endeavors to prioritize the advancement of 3D zinc anodes and alloys, aiming to fine-tune their composition and electrochemical behavior across a spectrum of acidic and alkaline conditions. Considering these factors, the following suggestions and viewpoints are proposed:

1. When confronting issues extending beyond the emergence of dendritic structures, the formation of passive layers through corrosion, and the HER, zinc anodes need to exhibit exceptional performance attributes. These encompass strong reversibility in zinc-ion plating and stripping processes, significant retention of discharge capacity, enduring stability throughout long-lasting cycles, and reliable functioning across varied environmental conditions. Furthermore, giving priority to reducing environmental effects and optimizing cost-efficient production is paramount to enable widespread implementation on a large-scale.
2. When devising tactics to protect zinc anodes, researchers ought to broaden their attention beyond the anode material exclusively and consider additional relevant parameters. The crucial nature of the interface between the anode and electrolyte, along with optimization of the electrolyte, cannot be overstressed. The performance of the anode is profoundly influenced by the selection of electrolyte, demanding additional exploration into diverse electrolyte aspects. This encompasses studying their behavior under real-time conditions, composition, their compatibility with electrodes, and practical feasibility.
3. Much remains to be comprehended at the mechanistic level regarding the malfunction mechanisms of zinc anodes. Through the utilization of in situ characterizations and theoretical simulations, researchers can attain a thorough comprehension of the reaction and transport mechanisms taking place at the surface of zinc anodes, thereby achieving a comprehensive insight. The fusion of these inquiries holds the capacity to offer valuable insights and design guidelines for tailored solutions.

It is worth mentioning that passivation, caused by the deposition of ZnS on the anode throughout battery operation, can severely hamper battery performance. Methodologies to alleviate this concern encompass the implementing anode materials or applying coating systems specifically designed to prevent the deposition of ZnS. As an illustration, employing anodes featuring augmented surface area or possessing unique surface chemistry can assist in better regulating the deposition of ZnS [29, 44]. An alternative methodology involves the inclusion of additive into the electrolyte aimed at preventing or reducing the generation of ZnS on the anode. These additives possess the capability to alter the attributes of the electrolyte, consequently reducing the probability of ZnS deposition or enhancing its dissolution [13, 15, 31]. In conclusion, successfully tackling the hurdles associated with Zn–S batteries demands a holistic methodology that places equal emphasis on advancements in both materials and designs. Improving cathode kinetics to boost sulfur utilization and proficiently regulating ZnS deposition on the anode are pivotal measures in the synthesizing of Zn–S batteries that exhibit both enhanced efficiency and prolonged durability.

References

1. Huang, S., et al., Chemistry (2019) 25
2. M. Zschornak et al., Phys. Sci. Rev. **3**(11), 20170111 (2018)
3. D. Liu et al., Nano Energy **101**, 107474 (2022)
4. A. Amiri et al., J. Mater. Chem. A **11**(20), 10788 (2023)
5. W. Yang et al., Joule **4**(7), 1557 (2020)
6. Y. Yan et al., Adv. Mater. **35**(18), 2211673 (2023)
7. G. Chang et al., Chem. Eng. J. **457**, 141083 (2023)
8. M. Cui et al., ACS Appl. Mater. Interfaces **13**(46), 54981 (2021)
9. A. Amiri et al., Mater. Today Phys. **24**, 100654 (2022)
10. G. Liang, C. Zhi, Nat. Nanotechnol. **16**(8), 854 (2021)
11. Shahali, H., et al., Energy Storage Materials (2023), 103130
12. B.-W. Zhang et al., Mater. Lab **1**(1), 210002 (2022)
13. L.-W. Luo et al., Chem. Commun. **57**(77), 9918 (2021)
14. W. Li et al., Adv. Sci. **7**(23), 2000761 (2020)
15. Z. Xu et al., Chem. Commun. **58**(58), 8145 (2022)
16. N. Zhang et al., J. Am. Chem. Soc. **138**(39), 12894 (2016)
17. C. Feng et al., J. Mater. Chem. A **11**(34), 18029 (2023)
18. Y. Guo et al., Small, 2207133 (2023)
19. T. Zhang et al., Energy Storage Mater. **33**, 181 (2020)
20. S. Licht, J. Electrochem. Soc. **134**(9), 2137 (1987)
21. R.D. Rauh et al., J. Inorg. Nucl. Chem. **39**(10), 1761 (1977)
22. A. Manthiram et al., Adv. Mater. **27**(12), 1980 (2015)
23. M.M. Gross, A. Manthiram, ACS Appl. Mater. Interfaces **10**(13), 10612 (2018)
24. D. Chao et al., Sci. Adv. **6**(21), eaba4098 (2020)
25. Q. Pang et al., Nat. Commun. **5**(1), 4759 (2014)
26. G. Zhou et al., ACS Nano **7**(6), 5367 (2013)
27. L. Zhang et al., Phys. Chem. Chem. Phys. **14**(39), 13670 (2012)
28. C. Zhang et al., Chem. Commun. **54**(100), 14097 (2018)
29. Y. Zhao et al., Adv. Mater. **32**(32), 2003070 (2020)
30. P. Cai et al., Adv. Energy Mater. **13**(28), 2301279 (2023)
31. J. Liu et al., J. Am. Chem. Soc. **145**(9), 5384 (2023)
32. H. Zhang et al., ACS Nano **16**(5), 7344 (2021)
33. A. Amiri et al., ACS Nano **17**(2), 1217 (2022)
34. W. Wu et al., Energy Environ. Sci. **16**(10), 4326 (2023)
35. L. Ma et al., Adv. Mater. **32**(14), 1908121 (2020)
36. Z. Zhao et al., Energy Environ. Sci. **12**(6), 1938 (2019)
37. A.P. Abbott et al., Electrochim. Acta **56**(14), 5272 (2011)
38. W. Kao-ian et al., J. Electrochem. Soc. **166**(6), A1063 (2019)
39. C. Zhang et al., Acc. Chem. Res. **53**(8), 1648 (2020)

40. J. Huang et al., J. Chromatogr. A **1598**, 1 (2019)
41. A.P. Abbott et al., Chem. Commun. **19**, 2010 (2001)
42. H. Qiu et al., Nat. Commun. **10**(1), 5374 (2019)
43. W. Zhang et al., Adv. Func. Mater. **33**(11), 2210899 (2023)
44. M. Yang et al., Angew. Chem. **134**(42), e202212666 (2022)
45. M. Liu et al., Batteries Supercaps **6**(7), e202300145 (2023)
46. T. Zhou et al., Mater. Today Energy **27**, 101025 (2022)
47. J. Li et al., Mater. Horizons (2023)
48. B. Yong et al., Adv. Energy Mater. **10**(45), 2002354 (2020)
49. G. Fang et al., ACS Energy Lett. **3**(10), 2480 (2018)
50. W. Lu et al., Chemsuschem **11**(23), 3996 (2018)
51. M. Wang et al., Adv. Energy Mater. **11**(5), 2002904 (2021)
52. J. Zhao et al., Nano Energy **57**, 625 (2019)
53. L. Ma et al., Energy Environ. Mater. **3**(4), 516 (2020)
54. Z. Kang et al., ACS Sustain. Chem. Eng. **7**(3), 3364 (2019)
55. Y. Hu et al., Mater. Today Energy **19**, 100608 (2021)
56. E. Foreman et al., Adv. Sustain. Syst. **1**(11), 1700061 (2017)
57. C. Li et al., Chem. Eng. J. **379**, 122248 (2020)
58. Y. Yu et al., Adv. Sustain. Syst. **4**(9), 2000082 (2020)
59. Y. An et al., ACS Nano **15**(9), 15259 (2021)
60. H. Zhang et al., Nat. Commun. **12**(1), 14 (2021)
61. D. Kundu et al., Nat. Energy **1**(10), 1 (2016)
62. L. Ma et al., Nat. Energy **5**(10), 743 (2020)
63. Y. Yang et al., Adv. Mater. **33**(11), 2007388 (2021)
64. N. Zhang et al., Angew. Chem. Int. Ed. **60**(6), 2861 (2021)
65. Z. Cai et al., J. Am. Chem. Soc. **143**(8), 3143 (2021)
66. Z. Cai et al., Energy Storage Mater. **27**, 205 (2020)
67. C. Liu et al., ACS Energy Lett. **6**(3), 1015 (2021)
68. Q. Zhang et al., Nat. Commun. **11**(1), 4463 (2020)
69. Y. Tang et al., Energy Storage Mater. **27**, 109 (2020)
70. Y. An et al., Adv. Func. Mater. **31**(26), 2101886 (2021)
71. D. Li et al., Angew. Chem. Int. Ed. **60**(23), 13035 (2021)
72. L. Cao et al., Angew. Chem. **132**(43), 19454 (2020)
73. Z. Chen et al., Energy Environ. Sci. **14**(6), 3492 (2021)
74. Q. Zhang et al., Angew. Chem. Int. Ed. **59**(32), 13180 (2020)
75. F. Xie et al., Adv. Energy Mater. **11**(9), 2003419 (2021)
76. Y. An et al., Chem. Eng. J. **400**, 125843 (2020)
77. Y. Zeng et al., Adv. Mater. **31**(36), 1903675 (2019)
78. X. Zhang et al., Energy Environ. Sci. **14**(5), 3120 (2021)
79. Y. Zhang et al., ACS Energy Lett. **6**(2), 404 (2021)
80. J.Y. Kim et al., Adv. Func. Mater. **30**(36), 2004210 (2020)
81. L. Kang et al., Adv. Energy Mater. **8**(25), 1801090 (2018)
82. C. Deng et al., Adv. Func. Mater. **30**(21), 2000599 (2020)
83. P. Liang et al., Adv. Func. Mater. **30**(13), 1908528 (2020)
84. J. Hao et al., Adv. Func. Mater. **30**(30), 2001263 (2020)
85. X. Liu et al., Advanced Science **7**(21), 2002173 (2020)
86. Y. Yin et al., Adv. Mater. **32**(6), 1906803 (2020)
87. Y. Tian et al., ACS Nano **13**(10), 11676 (2019)
88. Y. Tian et al., Energy Storage Mater. **41**, 343 (2021)
89. T. Zhang et al., Energy Environ. Sci. **9**(3), 1024 (2016)
90. H.D. Lim et al., Angew. Chem. **126**(15), 4007 (2014)
91. F. Wu et al., Adv. Mater. **27**(1), 101 (2015)
92. K.E. Sun et al., ACS Appl. Mater. Interfaces **9**(11), 9681 (2017)
93. X. Wu et al., J. Power Sour. **300**, 453 (2015)

94. J. Hao et al., Adv. Mater. **32**(34), 2003021 (2020)
95. L. Zhang et al., Adv. Func. Mater. **31**(26), 2100186 (2021)
96. C. Li et al., Energy Environ. Mater. **3**(2), 146 (2020)
97. Q. Zhang et al., Nat. Commun. **11**(1), 3961 (2020)
98. J. Liu et al., J. Am. Chem. Soc. **143**(38), 15475 (2021)
99. J. Cao et al., Adv. Energy Mater. **11**(29), 2101299 (2021)
100. W. Li et al., Adv. Func. Mater.Func. Mater. **31**(20), 2101237 (2021)
101. W. Li et al., Chem. Eng. J. **420**, 129920 (2021)

Chapter 4
Working Principles of Zinc–Sulfur Batteries

4.1 Charge and Discharge Processes

To propel the development and design of energy storage devices, it is imperative to unravel the nuanced charge–discharge behavior specific to Zn–S systems. This exploration into the fundamental mechanisms governing these electrochemical processes is pivotal for achieving a comprehensive understanding. By separating the intricate dynamics at play during charge and discharge cycles, researchers and engineers can glean invaluable insights. These insights, in turn, inform strategic decisions related to material choices, electrode configurations, and overall system architecture. The acquired knowledge serves as a cornerstone for the innovation and optimization of Zn–S-based energy storage solutions, paving the way for more efficient, durable, and advanced technologies in the realm of energy storage. This section of the book evaluates the processes of charging and discharging of Zn–S batteries.

The inquiry undertaken by Li et al. [1] scrutinized the processes unfolding on the zinc anode and the cathode containing sulfur in aqueous Zn–S batteries. Concerning Fig. 4.1a, b, showcasing ex situ XRD and Raman Spectra for the S@CNTs-50 cathode, reducing the battery charge to 0.3 V brings about a significant decrease in the intensity of the peak associated with S8, while concurrently, three prominent peaks manifest at diffraction angles (2θ) of 28°, 48°, and 56°, mirroring the characteristic peaks observed in ZnS. The inference drawn from this observation implies that sulfur undergoes consumption, leading to the formation of ZnS on the cathode surface throughout the discharge process. Upon reaching a discharge potential of 0.05 V, the steadfastness of the three ZnS peaks endures, whereas the complete disappearance of the S8 peak signifies the exhaustive consumption of sulfur.

In contrast, throughout the charging procedure, the resurgence of the sulfur distinctive peak at a potential of 1.2 V implies the synthesis of sulfur on the cathode surface. With the subsequent rise in potential, the intensity of sulfur peak amplifies, while the ZnS peaks experience a decline. At a potential of 1.6 V, denoting complete charge, a

A. Amiri et al., *The Zinc–Sulfur Battery*,
SpringerBriefs in Applied Sciences and Technology,
https://doi.org/10.1007/978-3-031-71491-7_4

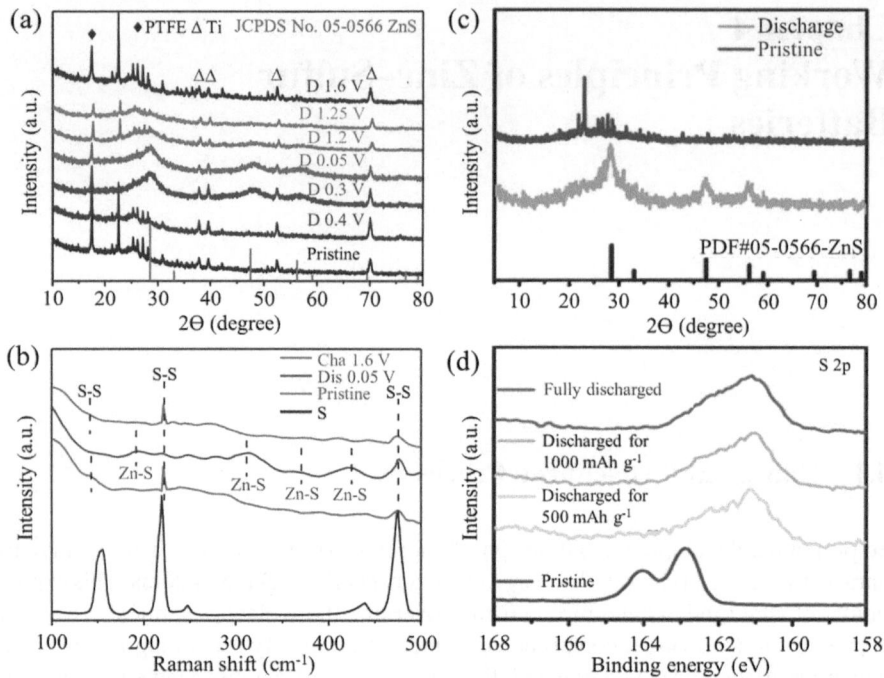

Fig. 4.1 **a** XRD profile and **b** Raman spectroscopy results acquired from the S@CNTs-50 cathode in a 1 M solution of $Zn(CH_3COO)_2$ at various charge and discharge phases [1]. **c** The XRD profile of the pristine and entirely discharged S@KB cathode. **d** The high-resolution XPS spectra represent the pristine and S@KB cathode in various charge states. Reproduced with permission from [1, 2]

conspicuous emergence of the S8 peak takes place, accompanied by the disappearance of ZnS signals. The Raman Spectra results collaborate the outcomes from XRD. Luo et al. [2] conducted analogous analyses on S@K. Examination of the S@KB electrode through ex situ XRD, as portrayed in Fig. 4.1c, indicates that at a discharge potential of 0.1 V, ZnS stands out as the singular discernible phase. This suggests the exhaustive consumption of sulfur throughout the discharge process. In Fig. 4.1d, detailed XPS spectra of the cathode at diverse charge stages are depicted. In its initial stage, the binding energies of sulfur (S 2p) are observed at 163 eV and 164 eV, aligning with the distinctive peaks of elemental sulfur. During the discharge phase, a reduction in the observed binding energies occurs, accompanied by the appearance of a wide peak, ranging from 160 eV to 163.5 eV. Concurrently, the intensity of this wide peak corresponding to ZnS's distinctive peak increases, signifying an enhanced synthesis of ZnS throughout the discharge processes.

4.2 Cycling Behavior

Cycling behavior, commonly referred to as cyclability, measures a battery's ability to sustain its performance and capacity through repeated charge and discharge cycles over time. It essentially reflects the battery's durability and longevity under continuous use.

The primary cause of deteriorating cycling behavior in Zn–S batteries is attributed to the formation of the by-product SO_4^{2-}. Chang et al. [3] conducted research demonstrating that incorporation of 2 g/L of TU redox mediator to a battery with a zinc anode, S@KB cathode, and $ZnSO_4$ electrolyte notably enhanced its electrochemical performance over 300 cycles. As shown in Fig. 3.10c, d, the battery with 2 g/L TU redox mediator exhibited minimal capacity loss, decreasing by only 0.11% per cycle at a current density of 5 A/g. In contrast, the specific capacity of the battery without the redox mediator significantly declined after just 20 cycles. This enhancement is attributed to the inhibition of SO_4^{2-} by-product formation.

The second factor contributing to the decline in cycling performance in Zn–S batteries is the formation of dendritic structures on the zinc electrode's surface. Liu et al. [4] found that adding 1 wt.% TUI to a battery composed of a ZnS@CF cathode, zinc anode, and 3 M $ZnSO_4$ aqueous electrolyte significantly improved cycle stability. As shown in Fig. 3.11c, there is only a slight change in charge and discharge capacities when the number of cycles increases to 5 and 10. This remarkable cycle stability of the Zn–S battery can be attributed to inclusion of 1 wt.% TUI in the 3 M $ZnSO_4$ electrolyte. Additionally, Fig. 3.12b demonstrates that incorporating the redox mediators TUI and ZnI_2 led to a consistent and high CE of approximately 100% after 100 cycles.

The cycling performance of Zn–S batteries is also compromised by irreversible side reactions. These reactions produce hydrogen and oxygen, and cause H_2S to form and escape from the electrolyte, which results in the depletion of sulfur within the electrolyte. As depicted in Fig. 3.13f, the cycling behavior of a battery configuration comprising a zinc anode, a high-strength CNF-S cathode, and a freeze-resistant electrolyte containing 1 M $ZN(CH_3COO)_2$, along with an EG freeze-proof agent and I_2 redox mediator, was assessed over 500 cycles at a current density of 1 A/g. The results indicate a notable decrease in capacity during the initial 300 cycles, attributed to irreversible side reactions that produce hydrogen, oxygen, and H_2S, leading to sulfur depletion. However, over the subsequent 200 cycles, the porous CNF-S cathode significantly contributes to stabilizing these reactions, resulting in only a 3% observed capacity decay. Furthermore, after 500 cycles, a CE approaching 100% is attained [5].

The cycling behavior of Zn–S batteries is significantly influenced by the formation of by-products, dendritic structures on the zinc electrode, and irreversible side reactions. Applying modifications to the different parts of the battery, encompassing the anode, cathode, and electrolyte, has proven effective in mitigating these issues and leading to enhanced electrochemical performance and cycle stability. These improvements highlight the potential for further optimizing Zn–S battery composition and configurations to achieve greater durability and longevity. With ongoing research and development, Zn–S batteries could become a more viable and efficient option for energy storage applications, contributing to the advancement of sustainable energy solutions.

4.3 Capacity and Efficiency Considerations

As mentioned earlier, Zn–S batteries have garnered significant attention in recent years due to their high theoretical energy density, low cost, and environmental friendliness. However, achieving optimal performance in terms of capacity and efficiency remains a critical challenge. Capacity, which refers to the amount of charge a battery can store and deliver, and efficiency, particularly CE, which measures the ratio of charge output to charge input during cycling, are pivotal parameters that determine the practical viability of Zn–S batteries. Factors such as the formation of by-products, dendritic growth on the zinc anode, and irreversible side reactions play a crucial role in influencing these parameters. Addressing these issues through strategic modifications to the anode, cathode, and electrolyte can significantly enhance the overall electrochemical performance of Zn–S batteries. This section of the book outlines the key considerations for improving the capacity and efficiency of Zn–S batteries, setting the stage for a deeper exploration of advanced strategies to overcome existing limitations.

To the best of our knowledge, considerable research has been conducted on Zn–S batteries aimed at improving their efficiency and capacity. One study focused on a battery composed of a zinc foil anode, an S@CMK-3 cathode, and a 3 M Zn(OTF)$_2$ + I$_2$ electrolyte. This battery exhibited a maximum capacity of 788 mAh/g$_S$ and a capacity retention of 65% at a current density of 1 A/g after 100 cycles [6]. Another study modified the cathode and electrolyte to S@KB and 1 M ZnCl$_2$ + DES, respectively, and included a LiCl$_2$ + AN redox mediator. This configuration resulted in a maximum capacity of 846 mAh/g$_S$ and a capacity retention of 38% at a current density of 1 A/g after 400 cycles [7]. Additionally, replacing the cathode and electrolyte in the Zn–S batteries with S@HCS and 2 M Zn(OTF)$_2$ + G$_4$ + I$_2$ delivered a maximum capacity of 937 mAh/g and a capacity retention of 80% at a current density of 1.7 A/g following 130 charge and discharge cycles [8]. Gross et al. [9] developed a Zn–S battery benefiting from 0.5M NaOH (Anolyte) and 0.1M Na$_2$S$_4$ + 0.1M NaOH as the electrolyte, along with a CoS@SS cathode. This developed battery provided a maximum capacity of 966 mAh/g$_{Na2S4}$ and a capacity retention of 85% after 50 cycles.

As shown in Table 3.1, various researchers [1, 2, 4, 8, 10–16] have achieved significant improvements in capacity and efficiency by modifying the electrolytes and cathodes in different Zn–S battery designs. However, the most efficient Zn–S batteries have been developed by Chang et al. [3] and Cai et al. [17]. Chang et al. [3] employed zinc foil anode, S@KB cathode, 2M ZnSO$_4$ electrolyte, and 2 g/L TU as a redox mediator. Their Zn–S battery demonstrated an impressive maximum capacity of 1,740 mAh/g$_S$ and a capacity retention of 51% at a current density of 5 A/g following 300 cycles. The most efficient electrochemical behavior so far has been demonstrated by Cai et al. [17]. They designed a Zn–S battery using an aqueous alkali/acid hybrid three-electrolyte configuration, an S/C cathode, and a zinc foil anode. This Zn–S battery achieved a maximum capacity of 2,250 mAh/g and maintained 86.5% of its capacity at a current density of 5 A/g over 400 charge and discharge cycles.

In conclusion, while Zn–S batteries hold great promise due to their inherent advantages, significant advancements are necessary to realize their full potential. The extensive research and strategic modifications to battery components, encompassing the anode, cathode, electrolyte, and redox mediator, have led to notable improvements in capacity and efficiency. The studies by Chang et al. [3] and Cai et al. [17] exemplify the progress that can be achieved through innovative design and optimization. As the field continues to evolve, further exploration and refinement of these strategies will be crucial in overcoming existing challenges and enhancing the practical viability of Zn–S batteries for widespread energy storage applications.

References

1. W. Li et al., Adv. Sci. **7**(23), 2000761 (2020)
2. L.-W. Luo et al., Chem. Commun. **57**(77), 9918 (2021)
3. G. Chang et al., Chem. Eng. J. **457**, 141083 (2023)
4. D. Liu et al., Nano Energy **101**, 107474 (2022)
5. A. Amiri et al., ACS Nano **17**(2), 1217 (2022)
6. Z. Xu et al., Chem. Commun. **58**(58), 8145 (2022)
7. M. Cui et al., ACS Appl. Mater. Interfaces **13**(46), 54981 (2021)
8. M. Yang et al., Angew. Chem. Int. Ed. **61**(42), e202212666 (2022)
9. M.M. Gross, A. Manthiram, ACS Appl. Mater. Interfaces **10**(13), 10612 (2018)
10. T. Zhou et al., Mater. Today Energy **27**, 101025 (2022)
11. W. Zhang et al., Adv. Func. Mater. **33**(11), 2210899 (2023)
12. Y. Zhao et al., Adv. Mater. **32**(32), 2003070 (2020)
13. H. Zhang et al., ACS Nano **16**(5), 7344 (2021)
14. J. Liu et al., J. Am. Chem. Soc. **145**(9), 5384 (2023)
15. Y. Guo et al., Small **19**(29), 2207133 (2023)
16. Y. Li et al., Mater. Horizons (2023)
17. P. Cai et al., Adv. Energy Mater. **13**(28), 2301279 (2023)

Chapter 5
Performance Characteristics and Challenges

5.1 Energy Density

Zn–S batteries have emerged as a promising candidate in the realm of energy storage technologies, primarily due to their remarkable energy density. Energy density, a critical parameter for evaluating battery performance, refers to the amount of energy a battery can store relative to its weight or volume. Zn–S batteries boast a high theoretical energy density, which makes them highly attractive for applications requiring efficient and compact energy storage solutions.

As mentioned earlier, the high energy density of Zn–S batteries stems from the intrinsic properties of sulfur and zinc. Sulfur, being abundant and low cost, offers a high theoretical capacity, while zinc provides excellent stability and safety profiles. Together, these elements create a synergy that results in batteries capable of storing significant amounts of energy in relatively small and lightweight packages. This characteristic is particularly beneficial for portable electronic devices, electric vehicles, and large-scale ESSs, where maximizing energy storage while minimizing weight and size is crucial.

However, realizing the full potential of Zn–S batteries' energy density in practical applications remains challenging. Issues such as the formation of insulating by-products, dendritic growth on the zinc anode, and polysulfide shuttle effects can negatively impact the energy density and overall performance of the batteries. Addressing these challenges through innovative material design and advanced engineering strategies is essential to harness the high energy density of Zn–S batteries fully.

This part of the book provides an overview of energy density in Zn–S batteries and the ongoing efforts to overcome existing barriers to their practical implementation.

© The Author(s), under exclusive license to Springer Nature Switzerland AG 2024   103
A. Amiri et al., *The Zinc–Sulfur Battery*,
SpringerBriefs in Applied Sciences and Technology,
https://doi.org/10.1007/978-3-031-71491-7_5

5.2   Environmental Impact

Certain metals and non-metals utilized in the production of battery can pose potential threats to human health through various exposure routes, including inhalation, skin or eye contact, ingestion, and injection. To illustrate, lead is typically absorbed by humans through ingestion, inhalation, and dermal absorption [1, 2], while cadmium can be absorbed through both ingestion and inhalation [3–5]. Mercury poses risks through inhalation, ingestion, and skin contact [6, 7]. Mousavi et al. [8] highlighted the harmful effects of lead, cadmium, mercury, arsenic, and chromium on human health, emphasizing that heavy metals, in general, pose risks to both public health and the environment [3, 9, 10].

Metal and their compounds enter soil, groundwater, and surface waters through various routes, including mining and industrial activities. Water sources are adversely impacted by landfills and tailings ponds, while the atmosphere is affected by the release of dust or evaporation from processes such as waste burning during recycling. The handling of waste materials from battery production and recycling poses a pressing and expanding issue for public health. This is attributed to their toxic properties, extensive distribution, and prolonged persistence in the environment [11]. For example, recycling processes for Pb-A batteries can release lead fumes and particles into the air [12, 13]. In the case of Li-ion batteries, they generate approximately $70 \, kg \, kW^{-1}h^{-1}$ of carbon dioxide (CO_2) [14]. It becomes crucial to consider CO_2 emissions across various stages, including mining, transportation, manufacturing processes, and recycling pathways, when conducting a holistic environmental assessment of batteries.

As previously noted, batteries are manufactured from a diverse array of materials, encompassing metals, non-metals, plastics, paper (or paperboard), and electrolytes. The environmentally conscious management of battery disposal, including collection, treatment, recycling, and burial, holds significant importance. The widespread apprehension surrounding used battery disposal stems from the inherent hazards associated with metallic waste, which incurs substantial costs for safe disposal [15].

5.3   Safety Consideration

As mentioned earlier, batteries are essential in our daily lives, yet their chemical makeup can pose significant dangers to humans and the environment. Improper use and exposure to extreme conditions like high and low temperatures can lead to cell fractures, releasing harmful liquids and gases. To mitigate fire and contamination risks, strict safety protocols must be adhered to in the manufacturing, storage, usage, and recycling of batteries. This scoping review highlights crucial safety, health, and environmental considerations for all batteries, specifically Zn–S batteries [16].

Regarding chemical stability and reactivity, the safety hazards of Zn–S batteries' two main constituents—sulfur and zinc—must be assessed individually. Elemental sulfur is relatively stable but can become reactive under certain conditions, such as high temperatures or in the presence of strong oxidizers. Similarly, zinc is generally stable but can corrode when exposed to moisture and certain chemicals. Thermal management is critical to prevent overheating, which could lead to thermal runaway and potentially hazardous situations, as both zinc and sulfur can participate in exothermic reactions that generate heat.

Additionally, electrolytes in Zn–S batteries—whether aqueous, gel, or organic—can be corrosive and require specific safety measures. Handling and disposal of these electrolytes should follow safety guidelines to avoid skin and eye contact or environmental contamination. This necessitates the use of appropriate personal protective equipment (PPE) and materials resistant to corrosion. Therefore, proper storage and handling are essential to prevent degradation and ensure safety.

Sulfur and zinc compounds share similar safety considerations. Sulfur can form various compounds, some of which, like hydrogen sulfide (H_2S), are toxic and pose inhalation hazards. Although elemental sulfur is not highly toxic, measures should be taken to prevent the formation of toxic by-products. Similarly, zinc and its compounds are generally considered low in toxicity, but they can cause irritation to the skin, eyes, and respiratory tract. Ingestion of large amounts can lead to zinc toxicity, which may result in nausea, vomiting, and other symptoms.

Both zinc and sulfur are recyclable, but the recycling process must be handled carefully to prevent environmental contamination. Proper disposal protocols should be in place to manage used batteries and prevent the release of harmful substances into the environment. Disposal of Zn–S batteries must comply with local hazardous waste regulations to avoid leaching harmful chemicals into the soil and water. It is important of not disposing these batteries in regular trash.

From fire and explosion perspective, sulfur is combustible and can ignite at temperatures above 190 °C (374 °F). Fire-related incidents occurring in battery manufacturing and recycling facilities pose significant dangers, leading to both direct and indirect consequences. Table 5.1 provides an overview of prominent fire emergencies documented within these plants between 2010 and 2018 [16]. Burning sulfur produces sulfur dioxide (SO_2), a toxic gas. Additionally, Zn–S batteries can produce hydrogen gas during charging or under certain conditions, which is highly flammable and poses explosion risks. Therefore, it is crucial to store sulfur and Zn–S batteries in cool, dry, and well-ventilated areas, away from direct sunlight and sources of heat and ignition. Equipment storage, production, and usage areas should be equipped with appropriate fire suppression systems, such as Class D fire extinguishers for metal fires. Additionally, train personnel in emergency response procedures specific to battery-related incidents to ensure prompt and effective action in case of any fire or explosion.

For operational safety, it is essential to follow manufacturer guidelines for charging and discharging to avoid conditions that could lead to overheating, overcharging, or deep discharging, all of which can compromise battery safety. Zn–S batteries must also be protected from physical damage, such as punctures, crushing, or impacts, which could lead to short circuits, leakage of electrolytes, or exposure of reactive materials.

Table 5.1 Significant fire emergencies at battery production and recycling facilities between 2010 and 2018. Reproduced with permission from [16]

Name and type of plants	City, country	Burning substances	Possible cause	Environmental consequences and contingency strategies	Year of incidents
Olympia Group, sunlight systems, battery manufacturing plant	Xanthi, Greece	Polymers, acids, and other combustible substances	Battery charger causing a short-circuit	Limited environmental repercussions occurred. An evacuation directive was issued, with certain residents advised to remain indoors	2018
Quallion LLC, battery manufacturing firm	Los Angeles, USA	LIBs	Unknown	Limited environmental repercussions occurred. No evacuation directive was issued	2018
Samsung SDI, battery manufacturing facility	Tianjin, China	Wastes from LIBs and deficient units	Disposed defective batteries	Unknown	2017
Amara Raja Batteries Limited, storage battery manufacturing company	Andhra Pradesh, India	Polymers, acids, and other combustible substances	Electrical short-circuit	Unknown	2017
Inmetco, battery recycling plant	Ellwood City, USA	Unknown	Unknown	Limited environmental repercussions occurred. An evacuation directive was issued, with certain residents advised to remain indoors	2015
Metair, first national battery, battery manufacturer and distributor	East London, South Africa	Whole facility	Short circuit within a battery or ignition caused by a damaged connector line	Unknown	2014

(continued)

Table 5.1 (continued)

Name and type of plants	City, country	Burning substances	Possible cause	Environmental consequences and contingency strategies	Year of incidents
Inmetco, Battery recycling plant	Ellwood City, USA	Unknown	Unknown	Unknown	2012
Metair, First national battery, battery manufacturer and distributor	Benoni, South Africa	Half the facility	Unknown	Unknown	2011
Unknown, battery factory	Suixi, China	Unknown	Unknown	Unknown	2011
Gould, battery plant	West Shreveport, USA	Unknown	Unknown	Limited environmental repercussions occurred. No evacuation directive was issued	2011
Toxco, lithium battery recycling facility	Trail, Canada	Unknown	Unknown	Limited environmental repercussions occurred. No evacuation directive was issued	2010

References

1. J.W. Cherrie et al., Ann. Occup. Hyg. **50**(7), 693 (2006)
2. A. Ara, J.A. Usmani, Interdiscip. Toxicol. **8**(2), 55 (2015)
3. M. Jaishankar et al., Interdiscip. Toxicol. **7**(2), 60 (2014)
4. A. Bernard, Indian J. Med. Res. **128**(4), 557 (2008)
5. P.B. Tchounwou et al., Molecular, clinical and environmental toxicology: volume 3: environmental toxicology, **133**
6. R.A. Bernhoft, J. Environ. Publ. Health **2012** (2012)
7. L. Keith et al., *Academica Press* (Burlington, USA, 2007)
8. M. Balali-Mood et al., Int. J. Occup. Environ. Med. **4**(2) (2013)
9. M.V. Gallegos et al., Waste Manage. **33**(6), 1483 (2013)
10. J.A. Guevara-García, V. Montiel-Corona, J. Environ. Manage. **95**, S154 (2012)
11. Y. Li, G. Xi, J. Hazard. Mater. **127**(1–3), 244 (2005)
12. B. Schaddelee-Scholten, J. Tempowski, United Nations Environment Programme on Safe Management of Used Lead Acid Batteries (2017)
13. T. PA, G.M. Eco, STRATEGIC PARTNERSHIP FOR THE MEDITERRANEAN SEA LARGE MARINE ECOSYSTEM. pastrategic
14. M. Armand, J.-M. Tarascon, Nature **451**(7179), 652 (2008)
15. E. Sayilgan et al., Hydrometallurgy **97**(3–4), 158 (2009)
16. S. Schismenos et al., Saf. Sci. **140**, 105290 (2021)

Chapter 6
Comparative Analysis with Other Battery Technologies

The significance of rechargeable batteries, known as secondary batteries, lies in their critical role in advancing the broad integration of renewable energy into our power grid and in everyday scenarios [1]. As previously noted, the past 20 years have seen substantial advancements in secondary batteries, driven by the goals of supporting sustainable development, promoting environmental friendliness, and guaranteeing energy stability. To meet these objectives, batteries need to deliver high electrochemical performance, encompassing efficiency, energy density, and power density, while also being affordable, durable, safe, and eco-friendly. Currently, LIBs are regarded as one of the top battery technologies due to their superior energy density, which ranges from 100 Wh/kg to 900 Wh/kg, acceptable power density of up to approximately 50 kW/kg, cyclability of around 500–1,000 cycles, and low safety. However, concerns are increasing regarding their cost, safety, limited supplies of lithium and other raw materials, and the adverse environmental impacts associated with their growing utilization [2, 3]. The pursuit of an alternative battery technology is fueled by this situation, aiming to use raw materials abundant on Earth, which reflects in cost reduction, while employing environmentally friendly, non-flammable, non-toxic electrolytes. This technology also strives to maintain excellent electrochemical performance, encompassing energy density, power density, and cyclability [4, 5]. The distinctive characteristics of zinc, encompassing stability, safety, volumetric energy density of $5,855 \, \mathrm{mAh/cm^3}$, reversibility in aqueous environments, and affordability, make zinc-based batteries desirable for battery researchers.

The appealing characteristics mentioned above have ignited widespread research efforts on aqueous zinc-ion batteries (AZIBs) in recent years. Despite extensive research aimed at improving AZIBs' efficiency, the cathodes typically utilized in these batteries face hurdles, encompassing the dissolution of materials and the adverse effects caused by the interactions between Zn^{2+} ions and the crystal lattice of the base material [6]. Additionally, the storage capacity of these cathodes utilizing intercalation typically falls within the range of 50–300 mAh/g, limiting the total energy density associated with AZIBs. This is despite the potential for higher energy

© The Author(s), under exclusive license to Springer Nature Switzerland AG 2024
A. Amiri et al., *The Zinc–Sulfur Battery*,
SpringerBriefs in Applied Sciences and Technology,
https://doi.org/10.1007/978-3-031-71491-7_6

density considering the theoretical capacity of the zinc anode, which can reach up to 820 mAh/g [7, 8]. It is worth noting that the measured energy density for AZIBs typically falls between 50 and 200 Wh/kg. Similarly, power density ranges from 100 to 200 W/kg. Additionally, cyclability is reported to be between 500 and 1,000 cycles [9–11].

Conversion-based cathodes offer superior energy density and hold promise as a potential solution to address this issue. Expanding on this notion, significant research efforts are ongoing in the realm of metal-sulfur batteries incorporating organic electrolytes, encompassing Li–S batteries [8]. As illustrated in Fig. 1.1, the energy density of Li–S batteries can range from 300 to 4,000 Wh/kg, with a potential power density of up to approximately 80 kW/kg. According to Benveniste et al. [3], Li–S batteries exhibit a cyclability of 50–100 cycles and demonstrate high safety risks.

Sulfur provides numerous benefits akin to zinc, including its plentiful availability, economical cost (0.25 USD per kilogram), eco-friendliness, and impressive theoretical capacity of 1,675 mAh/g. Moreover, through integrating zinc and sulfur, Zn–S batteries are being developed as a sustainable and affordable energy storage solution, boasting a maximum energy density of 3,800 Wh/kg$_s$, outperforming many other battery technologies. Moreover, the power capacity of Zn–S batteries ranges between 50 and 100 kW/kg. In addition, the process of assembling cells for Zn–S batteries is uncomplicated and can be carried out in a regular, standard environment without the necessity of specialized equipment such as gloveboxes or specialized dry rooms. This simplicity contributes significantly to the cost-effectiveness of Zn–S batteries and enhances their safety and accessibility [12, 13].

In their investigation, Benveniste et al. [3] conducted a thorough assessment of the electrochemical performance and safety considerations across various battery categories. Lead-acid batteries were found to exhibit an energy density ranging between 30 and 40 Wh/kg, accompanied by a power density of approximately 180 W/kg, and a cycle life spanning from 200 to 2,000 cycles. In the case of nickel-metal hydride (Ni-MH) batteries, the energy density was observed to vary from 30 to 80 Wh/kg, with a relatively higher power density of around 1,800 W/kg and observed cyclability ranging between 500 and 2,000 cycles. Similarly, Ni–Cd batteries demonstrated an energy density ranging from 40 to 60 Wh/kg, with a power density of 140–180 W/kg and a cyclability of 500–2,000 cycles. Moreover, all three batteries are categorized as safe options for various applications.

Observing Fig. 1.1, we can assess the performance of several other common types of batteries, encompassing Na-ion, Li–O$_2$, and Al-ion batteries, in detail. Na-ion batteries, for instance, exhibit an energy density ranging from 10 to 130 Wh/kg and a power density spanning from 0.2 to 9 kW/kg. Li–O$_2$ batteries, on the other hand, demonstrate a significant energy density, ranging from 2,000 to 5,000 Wh/kg, with a maximum power density of approximately 0.6 kW/kg. Additionally, Al-ion batteries show an energy density of around 100 Wh/kg and a power density of about 10 kW/kg [13].

References

1. M.M. Biswas et al., (2013) Towards implementation of smart grid: an updated review on electrical energy storage systems. Scientific Research Publishing (2013)
2. J.B. Goodenough, Y. Kim, Chem. Mater. **22**(3), 587 (2010)
3. G. Benveniste et al., J. Environ. Manage. **226**, 1 (2018)
4. T. Ould Ely et al., Front. Energy Res. **7**, 71 (2019)
5. J. Liu et al., Green Energy Environ. **3**(1), 20 (2018)
6. L.E. Blanc et al., Joule **4**(4), 771 (2020)
7. X. Liu et al., J. Energy Chem. **61**, 104 (2021)
8. P. Mu et al., Energy Fuels **35**(3), 1966 (2021)
9. C. Xu et al., Angew. Chem. Int. Ed. **51**(4), 933 (2012)
10. W. Shi et al., Chemsuschem **14**(7), 1634 (2021)
11. L. Ou et al., Chemsuschem **15**(19), e202201184 (2022)
12. D. Patel, A.K. Sharma, Energy Fuels **37**(15), 10897 (2023)
13. H. Shahali et al., Energy Storage Mater., 103130 (2023)

Chapter 7
Future Direction and Outlook

7.1 Potential Advancements

While it is true that Zn–S batteries grapple with diverse hurdles encompassing their restricted cycling longevity and modest energy efficiency, they also possess numerous benefits and appealing attributes that highlight their potential real-world implementation. Certainly, the diminished energy efficiency of Zn–S batteries predominantly stems from several underlying parameters. Firstly, there is often an insufficiency of sulfur employment within the battery system, resulting in inefficiencies in energy conversion. Additionally, throughout the zincation procedure, the sulfur-based cathode tends to expand, leading to structural instability and decreased overall performance. The high possibility of HER occurring within the battery further diminishes its efficiency. Zn–S batteries also commonly exhibit low discharge voltage, which contributes to reduced energy output. This, coupled with significant polarization effects, further hampers the battery's performance. Furthermore, the repetitive transformation or interaction between sulfur and sulfide compounds during the battery's operation can induce deterioration and depletion of the active materials. This is exacerbated by the corrosion of the zinc anode and the formation of dendritic structures on the zinc anode surface, both of which impair the battery's cycling longevity.

Indeed, these parameters collectively contribute to the diminished energy efficiency and performance of Zn–S batteries, highlighting the pressing demand for considerable strides, innovations, and developments in the realm of Zn–S battery technology to tackle these hurdles effectively and enhance battery performance.

One of the key methodologies being pursued involved the exploration and development of advanced electrode materials. This encompasses research into sulfur nanostructures and the incorporation of conductive additives. These advanced materials are specifically designed to improve the electrical conductivity and enhance the structural resilience of cathodes that utilize sulfur. By enhancing conductivity and durability, these materials aim to tackle some of the key hurdles of Zn–S batteries.

© The Author(s), under exclusive license to Springer Nature Switzerland AG 2024   113
A. Amiri et al., *The Zinc–Sulfur Battery*,
SpringerBriefs in Applied Sciences and Technology,
https://doi.org/10.1007/978-3-031-71491-7_7

Furthermore, significant research efforts are underway to develop effective electrolytes and membrane designs. These advancements are focused on reducing the formation of polysulfide intermediates within the battery system. Polysulfide intermediates are known to degrade battery performance over time. By minimizing their formation, researchers aim to overcome existing challenges and pave the way for the development of more efficient and durable Zn–S batteries for various applications.

In addition, passivation, a phenomenon caused by the accumulation of ZnS on the anode during the operation of Zn–S batteries, poses a significant challenge to battery performance. This build-up of ZnS can obstruct the flow of ions and electrons within the battery, leading to decreased efficiency and capacity retention over time. One approach to tackle this hurdle involves the exploration of alternative anode materials that are less susceptible to passivation. This methodology involves extensive material characterization and optimization to identify suitable alternatives that maintain high conductivity and stability throughout the battery's lifespan. Another strategy focuses on the development of specialized coating systems designed to prevent the deposition of ZnS on the anode surface. These protective coatings act as a barrier, preventing the interaction between the anode material and sulfur compounds present in the electrolyte. This approach requires careful selection of coating materials and deposition techniques to ensure effective coverage and adhesion to the anode surface.

Indeed, the outlined suggestions serve as a comprehensive roadmap for potential advancements in Zn–S batteries. However, it is essential to acknowledge that further research and development efforts are imperative to refine these approaches and address the remaining challenges effectively. These endeavors will not only improve the energy efficiency and cycling durability of Zn–S batteries but also pave the way for their widespread adoption across various applications. Continued dedication to research and development in these areas is essential to unlock the full potential of Zn–S battery technology and propel it toward commercial viability and market acceptance.

7.2 Market Trends

Upon scrutinizing prevailing market trends, it is evident that significant advancements are underway in diverse sectors, encompassing electric vehicles (EVs), unmanned aerial vehicles (UAVs), all-electric airplanes (AEAs), electronic accessories, and medical instruments. In light of these developments, market demands for superior energy storage devices are increasing. These devices must not only offer high performance but also possess qualities such as ductility, bendability, and wearability. Consequently, there is a critical need for further exploration and innovation, particularly concerning Zn–S batteries [1, 2].

Ongoing research efforts focused on refining electrolyte and electrode formulation, elucidating underlying mechanisms, and simplifying processing techniques. By advancing in these areas, we can facilitate the development of Zn–S batteries that are both ductile and high-performing, meeting the evolving needs of diverse applications in the market. A multitude of zinc-based batteries have already entered commercial production and are being employed across diverse sectors within the battery market. Ni–Zn batteries exhibit promising prospects for applications in massive grid energy storage as well as portable and ductile ESSs. Zn–Co batteries, with meticulously engineered electrode materials, demonstrate capabilities for use in EVs and hybrid EVs. Zinc–air and hybrid alkaline zinc batteries offer favorable opportunities for integration into EVs and portable ESSs. This is attributed to their significant energy density, affordability, and environmentally sustainable features [1].

Despite these promising examples of zinc-based batteries, there is a significant focus on improving Zn–S batteries to align with market trends due to their inherent features that make them particularly suitable for the market. Despite the gains mentioned in Sect. 1.2 "Historical Background," further research is needed to address issues such as electro-chemo-mechanics, temperature, and safety to fully assess their market potential. Challenges like fatigue, deformation, wear, cavitation, and the effects of cyclic loading on interfacial reactions and ionic transport still need to be resolved. As well as electrolytes, these issues could be addressed by fabricating multifunctional electrode material that serves as structural ESSs, enabling both load bearing and energy storage capabilities concurrently. This approach aims to achieve weight saving at a system level [2].

The objective is to develop more affordable batteries that can offer cost-effective grid storage and enable EVs to cover much longer distances on a single charge [3]. Worries about availability of essential battery materials such as cobalt and lithium are driving the exploration of alternatives to conventional lithium-ion chemistry [3]. To sum up, the outlook for Zn–S batteries appears optimistic, showcasing potential applications across diverse fields such as energy storage, wearable electronics [4], and air-compatible structural batteries [2]. Nonetheless, there remain obstacles to address, encompassing issues like zinc dendrite growth, corrosion passivation, the HER, and the decomposition of active materials. However, with persistent research and development endeavors, these obstacles can be tackled, laying the groundwork for the extensive utilization of Zn–S batteries in the future.

7.3 Challenges and Opportunities

While there have been considerable advancements in the appropriate design of battery elements including electrode materials and electrolytes, there are still numerous hurdles to overcome for Zn–S batteries. Currently, most research on these batteries is centered around the cathode and electrolytes. The goal is to identify optimal compositions and architectures that provide an appropriate electrochemical window and exceptional performance [1].

As an illustration, in the realm of alkaline rechargeable zinc-based batteries, there are several fundamental research hurdles. Concerning nickel-based cathodes, β-Ni(OH)$_2$ exhibits limited energy density, whereas α-Ni(OH)$_2$, which features a broader interlayer distance, encounters instability during the transition to the β phase.

Regarding Co-based cathode, although the Co_3O_4 cathode achieves a capacity of 162 mAh/g, it falls significantly short of its theoretical capacity of 446 mAh/g. Hence, additional investigation is required to attain complete reversibility of the Co_3O_4 cathode. As for air-based and hybrid cathodes, improved electrocatalysts and cathode architecture are necessary to optimize their performance [1].

The ongoing strides are aimed at improving the overall energy efficiency and cycling durability of cells by optimizing their redox reactions and minimizing the generation of dendritic structures. As an example, Cai et al. [5] conducted a recent investigation that unveiled an energy density of 3,800 Wh/kg in half-cell setups, exhibiting considerable progress. This establishes the battery as a strong contender against all alkaline batteries and possibly even high-energy–density batteries such as Li-ion, especially in the event of Li scarcity leading to considerable price increases. The economical nature of Zn–S batteries renders them highly appealing for utilization in grid storage purposes. The pathway to enabling the broad adoption of Zn–S batteries across various applications in the future involves tackling the present hurdles through continuous investigation and development strides. Additionally, Zn–S batteries may capitalize on their appropriateness for stationary energy storage purposes, owing to their chemical composition that offers comparative safety and biocompatibility, in comparison with alternative components currently available. The safety and affordability attributes of Zn–S batteries can additionally aid in improving their applicability for energy storage and power systems in remote or off-grid settings, particularly in regions where reliable energy sources are necessary for sustainable operation in isolated areas. Furthermore, these batteries offer a further notable benefit as they possess a capability to be incorporated into renewable energy systems, encompassing wind turbines or solar panels, owing to their ability to operate efficiently across a broad spectrum of temperature extremes, spanning from very low to high temperatures.

Despite the potential of Li–S batteries as advanced metal-sulfur ESSs, the scarcity of lithium has driven researchers to explore other options, such as Zn–S batteries. These Zn–S batteries are non-flammable, perform well in cryogenic conditions, and are 30% lighter than traditional alkaline batteries, positioning them as the next breakthrough in energy storage technology. Additionally, the Zn–S batteries make a compelling case as alternatives to ZIBs in various applications due to their low cost, environmental benefits, non-flammability, high theoretical capacity, superior performance, and light weight nature.

References

1. Y. Shi et al., Small **16**(23), 2000730 (2020)
2. A. Amiri et al., ACS Nano **17**(2), 1217 (2022)
3. C. Feng et al., J. Mater. Chem. A **11**(34), 18029 (2023)
4. A. Amiri et al., J. Mater. Chem. A **11**(20), 10788 (2023)
5. P. Cai et al., Adv. Energy Mater. **13**(28), 2301279 (2023)

Conclusion

Summary of Key Points

This book presented a comprehensive exploration of battery technology, delving into its unique insights, spanning an overview, historical context, and the significance and applications of Zn–S batteries. Following this, it delves into battery chemistry fundamentals, dissecting electrochemical principles and the reactions within all components of Zn–S batteries.

Moving forward, the text scrutinizes the materials and components employed in electrolytes, separators, cathodes, and anodes. It also considers prospects concerning electrolytes, cathodes, and anodes, with a detailed evaluation covering aqueous, hydrogel, and organic (non-polar) types of electrolytes, both with and without redox mediators, as expounded in Sect. 3.2.

Furthermore, the performance of Zn–S batteries is rigorously examined, highlighting the significant influence of electrolyte composition, zinc concentration, and the inclusion of redox mediators. The book proceeds with an in-depth analysis of the benefits and limitations of various electrolytes, exploring how functional additives and redox mediators can mitigate these limitations.

Section 3.3 shifts focus to the development of cathodes in Zn–S batteries, discussing a spectrum of carbon forms, ranging from spheres to fibers, and emphasizing the creation of HCS. It assesses the impact of these cathode modifications on the performance and longevity of Zn–S batteries, spanning applications from lab-scale prototypes to consumer-ready products. Additionally, within the same section, the book delves into the progressions and potential future directions concerning zinc anodes. Although zinc foil anodes have traditionally held sway, the text underscores pivotal considerations in opting for alternatives and delineates three separate classifications for such alternatives. This book further investigates the intrinsic obstacles linked with Zn–S batteries, organizing them based on distinct cell components, including electrolyte, anode, and cathode. Among the primary challenges are a restricted ESPW, formation and corrosion of zinc dendrites, sluggish kinetics at

A. Amiri et al., *The Zinc–Sulfur Battery*, SpringerBriefs in Applied Sciences and Technology, https://doi.org/10.1007/978-3-031-71491-7

the cathode, degradation of the cathode, and fading capacity, all of which are thoroughly addressed. Notably, an electrolyte-related challenge in Zn–S batteries is the occurrence of the HER. The electrolyte's composition, pH level, and ESPW surface as pivotal factors influencing the occurrence of the HER. This brief examination of hurdles sets the stage for the subsequent part of the review. Subsequently, this book outlines the studies conducted to confront the challenges related to electrolytes. It also explores alterations and advancements in electrolyte compositions, encompassing the incorporation of additives, adjustments in pH levels, and the investigation of alternative electrolyte configurations. Apart from addressing electrolyte concerns, researchers have directed their attention toward enhancing the design of cathodes and anodes. The books' review delves deeply into the studies aimed at overcoming challenges associated with both cathodes and anodes. More specifically, this book examines the diverse materials serving as hosts for cathodes in Zn–S batteries and how they influence battery performance. Features, benefits, and performance metrics of different cathode materials were thoroughly assessed to offer a comprehensive insight into the array of options and techniques for enhancing the performance of cathode in Zn–S batteries.

Chapter 3 of the book focus on an in-depth analysis of the electrochemical characteristics and effectiveness of Zn–S batteries. This chapter delves into crucial aspects such as charge and discharge behavior, cycling performance, energy density, capacity, and efficiency, while also considering environmental and safety considerations. Additionally, the book conducts a comparative study with other prevalent battery technologies such as LIBs and lead-acid batteries, thoroughly examining their respective pros and cons.

Ultimately, the conclusion extends its gaze toward the horizon of future possibilities, delving into potential advancements poised to shape the trajectory of Zn–S batteries. It further explores emerging market trends, illuminating the evolving landscape within which these batteries operate. Moreover, it navigates through the intricacies of challenges and opportunities, offering a panoramic view of the road ahead for Zn–S batteries in the realm of energy storage solutions.